Armin Ridinger

Towards ultracold polar 6Li40K molecules

Armin Ridinger

Towards ultracold polar 6Li40K molecules

Construction of an experimental apparatus, photoassociation of excited 6Li40K* molecules and a new manipulation method

Südwestdeutscher Verlag für Hochschulschriften

Impressum/Imprint (nur für Deutschland/only for Germany)
Bibliografische Information der Deutschen Nationalbibliothek: Die Deutsche Nationalbibliothek verzeichnet diese Publikation in der Deutschen Nationalbibliografie; detaillierte bibliografische Daten sind im Internet über http://dnb.d-nb.de abrufbar.
Alle in diesem Buch genannten Marken und Produktnamen unterliegen warenzeichen-, marken- oder patentrechtlichem Schutz bzw. sind Warenzeichen oder eingetragene Warenzeichen der jeweiligen Inhaber. Die Wiedergabe von Marken, Produktnamen, Gebrauchsnamen, Handelsnamen, Warenbezeichnungen u.s.w. in diesem Werk berechtigt auch ohne besondere Kennzeichnung nicht zu der Annahme, dass solche Namen im Sinne der Warenzeichen- und Markenschutzgesetzgebung als frei zu betrachten wären und daher von jedermann benutzt werden dürften.

Coverbild: www.ingimage.com

Verlag: Südwestdeutscher Verlag für Hochschulschriften GmbH & Co. KG
Dudweiler Landstr. 99, 66123 Saarbrücken, Deutschland
Telefon +49 681 37 20 271-1, Telefax +49 681 37 20 271-0
Email: info@svh-verlag.de

Approved by: Paris, Ecole Normale Supérieure, Diss., 2011

Herstellung in Deutschland:
Schaltungsdienst Lange o.H.G., Berlin
Books on Demand GmbH, Norderstedt
Reha GmbH, Saarbrücken
Amazon Distribution GmbH, Leipzig
ISBN: 978-3-8381-2895-5

Imprint (only for USA, GB)
Bibliographic information published by the Deutsche Nationalbibliothek: The Deutsche Nationalbibliothek lists this publication in the Deutsche Nationalbibliografie; detailed bibliographic data are available in the Internet at http://dnb.d-nb.de.
Any brand names and product names mentioned in this book are subject to trademark, brand or patent protection and are trademarks or registered trademarks of their respective holders. The use of brand names, product names, common names, trade names, product descriptions etc. even without a particular marking in this works is in no way to be construed to mean that such names may be regarded as unrestricted in respect of trademark and brand protection legislation and could thus be used by anyone.

Cover image: www.ingimage.com

Publisher: Südwestdeutscher Verlag für Hochschulschriften GmbH & Co. KG
Dudweiler Landstr. 99, 66123 Saarbrücken, Germany
Phone +49 681 37 20 271-1, Fax +49 681 37 20 271-0
Email: info@svh-verlag.de

Printed in the U.S.A.
Printed in the U.K. by (see last page)
ISBN: 978-3-8381-2895-5

Copyright © 2011 by the author and Südwestdeutscher Verlag für Hochschulschriften GmbH & Co. KG and licensors
All rights reserved. Saarbrücken 2011

A mon épouse, Maya.

Abstract

This book reports on the first successful steps towards the creation of ultracold polar diatomic molecules which are composed of two different atomic species of fermionic nature: ^{6}Li and ^{40}K. We have constructed a new experimental apparatus, which allows the preparation of the two-component atomic gas mixture in the quantum degenerate regime. The design, implementation and characterization of the apparatus are described in detail. We have realized a very large atom number dual-species magneto-optical trap and a magnetic transport of the two species over a large distance. We have used the atomic mixture to create, for the first time, excited heteronuclear ^{6}Li^{40}K* molecules by photoassociation. We have recorded and assigned photoassociation spectra for the most weakly bound states of seven excited molecular potentials and inferred the shape of the molecular potentials at long range. Our results pave the way for the production of ultracold bosonic ground-state ^{6}Li^{40}K molecules which are characterized by a large permanent electric dipole moment. Finally we have developed theoretically a novel method for the manipulation of quantum particles, which could be applied to polar ^{6}Li^{40}K molecules. This method consists in trapping the particles in a rapidly oscillating potential and to induce an instantaneous change of phase of the trapping potential (a phase hop). We show that the particle's mean motion can thus be significantly manipulated in a controlled fashion. This method has found a first application for Bose-Einstein condensates trapped in a time-averaged orbiting potential.

Zusammenfassung

In diesem Buch beschreiben wir die ersten erfolgreichen Schritte auf dem Weg zur Erzeugung ultrakalter polarer zweiatomiger Moleküle, welche sich aus zwei unterschiedlichen Atomsorten fermionischer Natur zusammensetzen: ^{6}Li und ^{40}K. Wir haben eine neue Versuchsanordnung aufgebaut, welche das zweikomponentige atomare Gasgemisch im quantenentarteten Zustand präparieren kann. Das Design, die Umsetzung und Charakterisierung der Versuchsanordnung werden detailliert beschrieben. Wir haben eine zwei-Spezies magneto-optische Falle mit großer Atomzahl und einen magnetischen Transport der beiden Atomsorten über eine große Wegstrecke verwirklicht. Wir haben das unter Ultrahochvakuum hergestellte Gasgemisch dazu benutzt um erstmals angeregte ^{6}Li^{40}K* Moleküle mittels Photoassoziation zu erzeugen. Wir haben Photoassoziationsspektren für die am schwächsten gebundenen Zustände von sieben angeregten Molekülpotenzialen aufgenommen und interpretiert und die Form der Potenziale im Bereich großer internuklearer Distanzen daraus abgeleitet. Unsere Ergebnisse ebnen den Weg für die Herstellung von ultrakalten bosonischen ^{6}Li^{40}K Molekülen im rovibrationellen elektronischen Grundzustand, welche durch ein großes permanentes Dipolmoment charakterisiert sind. Schließlich haben wir theoretisch eine neue Teilchenmanipulationsmethode entwickelt, die sich auf polare ^{6}Li^{40}K Moleküle anwenden lässt. Diese Methode besteht darin, die Teilchen in einem schnell oszillierenden Potenzial zu speichern und eine augenblickliche Änderung der Phase des Potenzials zu induzieren (eine Phasensprung). Wir zeigen, dass die zeitgemittelte Bewegung der Teilchen dadurch signifikant und auf kontrollierte Weise manipuliert werden kann. Diese auf beliebige Teilchen anwendbare Methode fand bereits eine erste Anwendung für Bose-Einstein Kondensate, die in einer sogenannten "TOP"-Magnetfalle gespeichert sind.

Résumé

Dans ce livre nous décrivons les premières étapes cruciales vers la formation, à très basse température, de molécules diatomiques dipolaires composées de deux espèces fermioniques: ^{6}Li et ^{40}K. Nous avons construit un nouveau dispositif expérimental qui permet la réalisation du mélange de gaz à deux composants dans le régime de dégénéréscence quantique. La mise en place et la caractérisation du dispositif sont décrites en détail. Nous avons réalisé un piège magnéto-optique à deux espèces avec un très grand nombre d'atomes, et un transport magnétique sur une grande distance. Les premières expériences avec le mélange atomique ont permis la première création de molécules hétéronucléaires excitées ^{6}Li^{40}K* par photoassociation. Nous avons enregistré et assigné des spectres de photoassociation pour les états les plus faiblement liés de sept potentiels moléculaires et nous en avons déduit la forme des potentiels à longues distances. Nos résultats ouvrent la voie vers la formation de molécules bosoniques ^{6}Li^{40}K ultra-froides dans leur état fondamental, caractérisé par un grand moment dipolaire électrique permanent. Sur le plan théorique, nous avons développé une nouvelle méthode pour la manipulation des particules quantiques, qui pourrait être appliquée aux molécules dipolaires ^{6}Li^{40}K. Cette méthode consiste à piéger les particules dans un potentiel oscillant rapidement et induire un changement instantané de phase du potentiel (un saut de phase). Nous montrons que le mouvement moyen des particules peut ainsi être manipulé de manière contrôlée. La méthode proposée a trouvé une première application pour les condensats de Bose-Einstein piégés à l'aide d'un piège magnétique du type "TOP".

Acknowledgments

The content of this book is based on my PhD thesis which I realized during the years 2007-2011 at the Laboratoire Kastler Brossel of the Ecole Normale Supérieure in Paris. Many people have been involved in the research projects which I describe here. I would like to thank in particular Christophe Salomon, Frédéric Chevy, Saptarishi Chaudhuri, Olivier Dulieu, Christoph Weiss, Thomas Salez, Ulrich Eismann, Nir Davidson, David Guéry-Odelin, David Wilkowski and Diogo Rio Fernandes for their support.

Contents

1 Introduction   1
    1.1 Quantum degenerate Fermi gases 2
    1.2 Mixtures of Fermi gases 5
    1.3 Polar Fermi-Fermi molecules 6
    1.4 Outline of this book 8

2 Construction of the experimental apparatus   11
    2.1 Design considerations 11
    2.2 Vacuum manifold 13
       2.2.1 Setup 13
       2.2.2 Assembly, pump down and bake out 15
    2.3 Laser systems 17
       2.3.1 Optics 18
       2.3.2 Diode lasers 20
       2.3.3 Saturated absorption spectroscopy 21
       2.3.4 Tapered amplifiers 21
    2.4 ^{6}Li Zeeman slower 24
       2.4.1 Principle of Zeeman-tuned slowing 24
       2.4.2 Oven 24
       2.4.3 Coil assembly 26
       2.4.4 Optics 27
    2.5 ^{40}K 2D-MOT 29
       2.5.1 Principle of a 2D-MOT 30
       2.5.2 Experimental setup 30
    2.6 ^{6}Li-^{40}K dual-species MOT 32
       2.6.1 Principle of a MOT 33
       2.6.2 Experimental setup 33
    2.7 Magnetic trapping 35
       2.7.1 Principle of magnetic trapping 36

		2.7.2	Transfer from the MOT to the magnetic quadrupole trap	39
		2.7.3	Magnetic transport	41
		2.7.4	Magnetic quadrupole trap of the final cell	44
		2.7.5	Optical plug	45
		2.7.6	Evaporative cooling	46
	2.8	Diagnostic tools		48
		2.8.1	Principle of absorption imaging	48
		2.8.2	Evaluation of absorption images	50
		2.8.3	Optical setup	52
		2.8.4	Practical aspects	53
		2.8.5	Auxiliary detection systems	55
		2.8.6	Experiment control and data acquisition	56
	2.9	Conclusion and outlook		58

3 Characterization of the experimental apparatus   61

	3.1	^{6}Li Zeeman slower		61
	3.2	^{40}K 2D-MOT		65
	3.3	^{6}Li-^{40}K dual-species MOT		72
		3.3.1	Single-species MOTs	73
		3.3.2	Heteronuclear Collisions in the dual-species MOT	78
	3.4	Transfer of the atoms into the magnetic trap		82
	3.5	Magnetic quadrupole trap		84
	3.6	Magnetic transport		87
	3.7	Conclusion		89

4 Photoassociation of heteronuclear ^{6}Li^{40}K molecules   93

	4.1	Introduction		93
		4.1.1	Principle of photoassociation	93
		4.1.2	Applications of ultracold photoassociation	95
		4.1.3	Photoassociation of LiK* compared to other dimers	98
		4.1.4	Detection techniques for photoassociation	100
		4.1.5	Molecular potentials	102
		4.1.6	Selection rules	107
		4.1.7	Rotational barriers for ultracold ground-state collisions	108
		4.1.8	The LeRoy-Bernstein formula	110
		4.1.9	Previous work on LiK	113
	4.2	Experimental results		114
		4.2.1	Experimental setup	115

		4.2.2 Optimization of the photoassociation signal 115

        4.2.2   Optimization of the photoassociation signal 115
        4.2.3   Photoassociation spectroscopy of ^{40}K$_2^*$ molecules 118
        4.2.4   Photoassociation spectroscopy of ^{6}Li^{40}K* molecules 125
    4.3   Conclusion 139

5 Particle motion in rapidly oscillating potentials   143

    5.1   Introduction 143
    5.2   Classical motion in a rapidly oscillating potential 146
        5.2.1   Time-independent description 146
        5.2.2   Coupling between the mean motion and the potential's phase . 148
        5.2.3   The effect of a phase hop 149
    5.3   Quantum motion in a rapidly oscillating potential 154
        5.3.1   Time-independent description 154
        5.3.2   The effect of a phase hop 157
        5.3.3   Numerical simulations 163
    5.4   Consistency between classical and quantum mechanical results 167
        5.4.1   Coherent states 167
        5.4.2   Effect of phase hop on a coherent mean-motion state 168
    5.5   Conclusion 170

6 Conclusion   173

A Determination of vapor pressure by light absorption   177

B Saturation spectroscopy of the violet $4S_{1/2} \to 5P_{3/2}$ transition of K   181

C Engineering drawings   191

    C.1   Octagonal cell 191
    C.2   Science cell 192
    C.3   Tapered amplifier support for potassium 193
    C.4   Tapered amplifier support for lithium 196
    C.5   2D-MOT vacuum parts 198

D Publications   201

Chapter 1

Introduction

Since the realization of the first Bose-Einstein condensate in 1995, the study of ultracold dilute quantum gases has evolved into one of the most active research fields in contemporary physics. The motivation to study these gases is given by their extremely low temperatures (typically 100 nK), which allow the observation of quantum mechanical phenomena at low densities, where a remarkable degree of experimental control exists. Held in free space by electromagnetic forces under ultrahigh vacuum, the gases can be studied in an extremely clean environment. It is possible to prepare and analyze them in specific potentials and atomic states and to control the interatomic interactions. Dilute quantum gases therefore represent ideal model systems with which quantum phenomena and many-body problems can be studied.

All particles can be divided into two distinct categories depending on their intrinsic angular momentum, the spin. If the spin is an integer multiple of $h/2\pi$ a particle is called a boson, if it is a half-integer multiple it is called a fermion. The resulting distinct properties become manifest when identical particles get so close to each other that they can no longer be distinguished because their wave packets overlap. Fermions then obey Fermi-Dirac statistics and underlie Pauli's exclusion principle, which forbids two or more of them to occupy the same quantum state of the confining potential. In contrary, bosons obey Bose-Einstein statistics at ultralow temperatures and tend to occupy the quantum state with the lowest energy and to form a Bose-Einstein condensate (BEC), in which all atoms can be described by a single macroscopic wave function. The coherence of the atoms in this state leads to the frictionless motion of the atoms called superfluidity.

The phenomenon of Bose-Einstein condensation has been predicted to occur in an ideal gas in 1925 by Einstein [1] based on preceding work of Bose [2]. The condensation is a phase transition which is a purely statistical phenomenon and does not require any interactions. Bose-Einstein condensation was first discovered in the form of superfluid liquid ^{4}He in 1938 [3, 4, 5]. BEC in a dilute gas has been realized for the first time in 1995 with weakly-interacting neutral bosonic alkali atoms [6, 7, 8]. The principal difficulty of achieving BEC in a gas was the realization of the required ultralow temperatures. These are associated with the low density, which is a prerequisite for avoiding cluster formation or solidification before BEC can occur. A comprehensive introduction to BEC and a review of the exciting experiments which have been carried out with them can be found in Refs. [9, 10, 11, 12]. The most spectacular results comprise the observation of macroscopic matter wave interference [13], the creation of vortices [14, 15], which unambiguously proofs the superfluid character of BECs, dark [16, 17] and bright solitons [18, 19] and the observation of the superfluid to Mott insulator transition [20].

Fermions behave fundamentally different from bosons at ultralow temperatures. Since they cannot occupy the same quantum state, non-interacting fermions at absolute zero simply fill up the lowest quantum states of the confining potential one by one up to the so-called Fermi energy—forming the so-called Fermi sea. The transition from the classical to the quantum degenerate Fermi gas is not accompanied by a phase transition but occurs gradually. In the presence of attractive interactions, however, fermions can undergo a phase transition to a superfluid state. For this to happen, the fermions need to form pairs which then behave like bosons. Fermionic superfluidity occurs in metals at ultralow temperatures which become superconducting, where the electrons, which are fermions, form a superfluid.

In the following we give a short review of the research which has been conducted in the past years with ultracold fermions. Then we present the scientific long-term goals of our experiment and their motivations. Finally we give an overview of the content of this book.

1.1 Quantum degenerate Fermi gases

A gas of fermions is called quantum degenerate when its temperature is much smaller than the so-called Fermi temperature $T_\mathrm{F} = E_\mathrm{F}/k_\mathrm{B}$, where E_F is the Fermi energy and k_B the Boltzmann constant. The first degenerate Fermi gas has been realized in 1999 with ^{40}K at JILA [21]. Since then, degenerate Fermi gases have been created also for ^{6}Li [22, 23], metastable ^{3}He* [24], the rare earth element ^{173}Yb [25] and the alkaline earth element ^{87}Sr [26, 27]. In early experiments the atoms could be cooled to about one quarter of the Fermi temperature. To date cooling to well below one tenth of T_F has been achieved [28, 29].

Cooling of Fermi gases is more difficult than for Bose gases. In order to achieve cooling of a gas to the lowest temperatures, a cooling technique called "evaporative

1.1. Quantum degenerate Fermi gases

cooling" needs to be applied. This method relies on rethermalization by elastic collisions. Typically atoms in an ultracold gas only collide via s-wave collisions, for which the colliding partners have no relative angular momentum. Due to Pauli's exclusion principle, these collisions are, however, forbidden for indistinguishable Fermi atoms. In order to make evaporative cooling work for fermions, the gas thus has to be either prepared in two different internal states, such that collisions between atoms in different states can take place, or it has to be brought in contact with an actively cooled gas, typically a Bose gas which is evaporatively cooled. In either case the experimental cooling procedure is more complicated than for bosons. Another reason why fermions are more difficult to cool is that the collision rate decreases when they become degenerate. This is because scattering into a low-lying momentum state requires this state to be empty, which becomes less and less probable with decreasing temperature [21, 30]. Furthermore particle losses due to inelastic collisions, which generally limit the evaporative cooling process, are more detrimental for degenerate Fermi gases, since they can create hole excitations deep in the Fermi sea [31, 32].

The first realization of a degenerate Fermi gas in 1999 [21] was achieved by evaporatively cooling a gas of magnetically trapped ^{40}K atoms in two different internal states. This single-species evaporative cooling method has also been applied to ^{6}Li [33, 34] and ^{173}Yb [25], for which the laser-cooled cloud was loaded into an optical trap in which the mixture of the lowest hyperfine states was prepared. Sympathetic cooling of fermions with bosons has been demonstrated for the following atomic combinations: ^{6}Li-^{7}Li [23, 22], ^{6}Li-^{23}Na [28], ^{6}Li-^{87}Rb [35], ^{40}K-^{87}Rb [36, 37, 38, 39, 40], ^{3}He*-^{4}He* [24], ^{6}Li-^{40}K-^{87}Rb [41] and ^{6}Li-^{40}K-^{41}K [42]. For the Fermi-Fermi mixture ^{6}Li-^{40}K a combination of both cooling methods has been demonstrated. The Innsbruck group [43] evaporatively cools ^{6}Li in an optical trap in two spin states, sympathetically cooling ^{40}K along. The Amsterdam group [44] evaporatively cools ^{40}K in a magnetic trap in two spin states, sympathetically cooling ^{6}Li along.

Early experiments on degenerate Fermi gases concentrated on the study of one-component systems, which are non-interacting and thus represent realizations of nearly ideal Fermi gases. Thermodynamic properties of these gases were studied. Their deviations from a classical gas, *i.e.* Pauli blocking of collisions at ultralow temperatures and the existence of the Fermi pressure were investigated [21, 45, 22]. Then, interest grew in studying Fermi gases with interactions. Since most of the properties of real-life materials are determined by the behavior of interacting electrons, which are fermions, such systems provide a strong connection between ultracold quantum gases and condensed-matter systems. An interacting Fermi gas needs to be composed of two components. At ultralow temperatures interactions can be characterized by the s-wave scattering length a, which is related to the s-wave collision cross-section via $\sigma = 4\pi a^2$. It was recognized that a type of scattering resonance, known as a Feshbach resonance [46, 47], could be exploited to change a and thus the strength of the interatomic interactions. Feshbach resonances occur if two colliding atoms couple resonantly to a bound molecular state. The scattering length can be tuned since the relative energy of the colliding atoms and of the bound state can be changed by applying, e.g., a magnetic field [48]. This technique was first established in 1998 [49] for bosons and it opened the door for the study of strongly interacting gases.

The first step of the study of interacting Fermi gases was thus the search and characterization of Feshbach resonances. Several of such resonances were found in ^{6}Li [50, 51, 52] and ^{40}K [53, 54, 55] which allowed the creation of strongly interacting Fermi gases. These findings provided the necessary tools to study the pairing which can lead to superfluidity in degenerate interacting Fermi gases.

The formation of pairs can happen in two different ways, depending on the interaction. When the interaction is strongly attractive, the fermions can form diatomic molecules, in which case they are tightly bound in position space. This superfluid state is referred to as the superfluid BEC. When the interaction is weakly attractive they can form so-called Cooper pairs [56]. In contrast to the diatomic molecules, Cooper pairs are very weakly bound, spread over a large volume and are defined via the strong correlations between the two components in momentum space. The binding to Cooper pairs is the type of binding which is observed in superconducting materials, where the electrons form such pairs to allow for frictionless charge flow, as explained by the Bardeen-Cooper-Schrieffer theory [57]. This superfluid state is referred to as the BCS superfluid.

After Feshbach resonances in fermionic gases had been investigated, they were exploited to create weakly bound molecules [58], which subsequently were shown to exhibit exceptionally large lifetimes in the vicinity of a Feshbach resonance even for large interactions [59, 34, 60, 61]. This is in contrast to bosonic systems, where large three-body losses occur for strong interactions. The long lifetime of the associated weakly bound molecules was shown to result from the Pauli exclusion principle, which prohibits three or more fermions to come sufficiently close to each other to induce vibrational relaxation in the molecule [62]. It allowed the realization of a BEC superfluid in 2003 [63, 64, 34] and subsequently also the realization of a BCS-type superfluid [65, 66]. The stability of strongly interacting fermion pairs then made the study of the crossover between both superfluid states possible [67, 68]. It was found that the crossover connects the two superfluid states smoothly across the strongly interacting regime.

The superfluid character of a fermionic condensate was unambiguously identified for the first time by the creation of vortices [69]. More recent experiments with strongly interacting ultracold Fermi gases comprise the study of fermionic mixtures with population imbalance and the observation of a phase separation between the paired and unpaired fermions [70, 71], the measurement of the speed of sound in a Fermi gas [72], the measurement of critical velocities [73], the realization of a Mott insulator of fermions [74, 75] and the direct measurement of the equation of state of a strongly interacting Fermi gas [29, 76, 77].

1.2 Mixtures of Fermi gases

In the field of ultracold Fermi gases the study of mixtures of two different fermionic species with different mass is gaining interest. Both theoretical and experimental aspects motivate this study. Such mixtures allow the study of fermionic pairing in the case of unmatched Fermi surfaces, which are due to the mass imbalance of the two species. Symmetric BCS pairing cannot occur and new quantum phases with different pairing mechanisms are predicted, such as the Fulde-Ferrell-Larkin-Ovchinnikov (FFLO) state [78, 79] or the breached pair state [80, 81]. Other phenomena such as a crystalline phase transition [82] and the formation of long-lived trimers [83] are predicted. Mixtures of two different fermionic species further allow the creation of polar molecules, which have a long-range dipole-dipole interaction [84, 85]. Two different atomic species yield additional tunable parameters, such as the mass imbalance and species-specific potentials. The mass-imbalance can be varied in an optical lattice, where the effective mass of each species depends on the optical lattice parameters. Species-specific potentials allow a more convenient way to study impurity physics such as Anderson localization [86] and the study of systems with mixed dimensions [87].

The mixture ^{6}Li-^{40}K is a prime candidate for these studies. ^{6}Li and ^{40}K are the only stable fermionic alkali isotopes and thus belong to the experimentally best-mastered class of atoms. Moreover, both species have bosonic isotopes which can also be used to create boson-fermion gases. Furthermore, the difference in the electronic structure of the two species is large leading to a large electric dipole moment for heteronuclear diatomic ^{6}Li^{40}K molecules (3.6 D) [88].

So far, all research groups working with mixtures of different fermionic species have chosen the mixture ^{6}Li-^{40}K [89, 43, 44, 42], sometimes with an additional third bosonic component. Quantum degeneracy in the mixture has been reported by three groups [41, 43, 42]. Interspecies Feshbach resonances have been observed and characterized [90, 91, 92] and weakly bound molecules have been created by a magnetic field sweep across a Feshbach resonance [93].

In order to use a Fermi gas to model other physical systems, universal behavior is required, *i.e.*, a regime in which the behavior of the system is entirely determined by the fermionic nature of the particles and the scattering length, and does not depend on the details of the scattering potential. Universal behavior for atoms can typically only be achieved when the system exhibits strong Feshbach resonances (*i.e.*, resonances which have a large width). It has so far only been achieved in the single species ensembles ^{6}Li [94] and ^{40}K. It was recently reported [92] that it may be possible to reach the universal regime for the ^{6}Li-^{40}K-mixture due to the existence of a 1.5 Gauss-wide Feshbach resonance. Recent experiments have already demonstrated strong interactions between ^{6}Li-^{40}K [95].

1.3 Polar Fermi-Fermi molecules

One major motivation for studying the mixture ^{6}Li-^{40}K is the formation of ultracold polar molecules. Polar molecules have a long-range, anisotropic dipole-dipole interaction and thus may provide access to qualitatively new quantum regimes which are inaccessible for neutral atoms [96, 97, 98]. They have been proposed to be used for the study of ultracold chemistry [99]. Furthermore, they are considered as excellent candidates for the realization of qubits for quantum computation [100, 101] or for fundamental tests like the measurement of the electron dipole moment [102, 103], the proton-to-electron mass ratio [104, 105] or the fine structure constant [106]. The values of these constants may drift monotonically over time, or vary periodically with the sun-earth distance in case of the existence of gravitational coupling [98]. In particular, polar molecules can be efficiently trapped and their interactions can be controlled by AC and DC electric fields. The alkali dimer LiK with its large dipole moment is a good candidate for these studies. The isotopomer ^{6}Li^{40}K, being composed of two fermions, forms a boson. Thus, evaporative cooling might be applicable to achieve Bose-Einstein condensation of polar molecules. Such a BEC would represent a quantum fluid of strongly and anisotropically interacting particles and thereby greatly enhance the scope for study and applications of collective quantum phenomena [107, 98].

While atoms are routinely laser cooled to ultracold temperatures, the complex internal structure of molecules makes this direct method difficult [108, 109]. The only molecule for which laser cooling could lead to microKelvin temperatures is SrF [110]. The methods which are used today in order to produce ultracold molecules can be divided in direct and indirect methods. In direct methods preexisting (typically hot) molecules are actively cooled, *e.g.*, by buffer gas cooling [111], Stark-slowing [112] or beam skimming with a guide [113]. These techniques currently allow reaching minimum

1.3. Polar Fermi-Fermi molecules

temperatures of $\sim 1\,\mathrm{mK}$, far too high to achieve molecular quantum degeneracy.

Indirect techniques yield access to much lower temperatures. They are based on the association of precooled atoms resulting in molecules which are translationally cold as well. The association can be either done using Feshbach resonances (magnetoassociation) or photoassociation techniques [114]. However, both methods typically create vibrationally excited molecules, which are unstable to collisions or do not have a significant dipole moment. In order to prepare deeply bound ground-state molecules, optical multi-color transfer schemes need to be employed. The transfer scheme with which the highest efficiencies could be obtained so far is the stimulated Raman adiabatic passage (STIRAP), based on a pair of continuous wave lasers (for an introduction see Ref. [115]). It was first introduced in the context of molecular spectroscopy in 1988 [116, 117] and applied to transfer magnetoassociated weakly bound molecules to very deeply bound molecules in the case of homonuclear Cs_2 [118] and even to the rovibrational ground state in the case of homonuclear Rb_2 [119] and heteronuclear RbK [85]. The transfer efficiencies in these experiments exceeded 80%. It is desirable to develop similar transfer schemes for $^6Li^{40}K$. Those require the precise knowledge of excited molecular potentials. One of the scientific contributions of this book is a study of the excited molecular potentials of $^6Li^{40}K$ via photoassociation spectroscopy, which will allow us to find pathways for the efficient optical transfer of atoms to ground-state molecules. The fermionic nature of both constituents of the molecule $^6Li^{40}K$ would be advantageous for the conversion process, since it leads to small losses in the association stage of the transfer scheme, which can be either done magnetically by Feshbach resonances [93] or optically by photoassociation, which, for molecules composed of bosons, typically induces significant loss.

Molecules in their absolute (rovibrational and spin) ground state are stable against inelastic collisions. However, reactive collisions can still occur. Such collisions have been studied recently for all pairs of alkali-metal dimers [120]. It has been shown that the molecule LiK is unstable against the exothermic atom exchange reaction LiK+LiK$\rightarrow$Li$_2$+K$_2$, which holds for any pair of dimers containing a Li atom. Reactive collisions of this kind have been observed also for the dimer RbK [85]. However, it has been shown to be possible to reduce these collisions by confining the molecules in a quasi-two-dimensional, pancake-like geometry, with the dipoles oriented along the tight confinement direction [121]. Two dimers can then approach each other only in a "side-by-side" collision, where the chemical reaction rate is reduced by the repulsive dipole-dipole interaction. Such trap geometries might thus also be required for the dimer $^6Li^{40}K$ when the study of dipolar gases is aimed for. On the other hand, the existence of reactive collisions allows the study of ultracold chemistry [99].

1.4 Outline of this book

The most time-consuming part of the described work consisted of the construction of a new experimental apparatus which prepares the mixture ^{6}Li-^{40}K for studies with the above-mentioned goals. When designing the apparatus special attention was paid to the creation of large atom numbers. The preparation of atomic clouds at ultracold temperatures typically requires several steps. First, the atoms have to be prepared in gaseous form inside an ultra-high vacuum environment. Then the atoms of the gas need to be captured, trapped and cooled. In our setup this step is efficiently realized by a magneto-optical trap (MOT), which is a dissipative trap employing an interplay between magnetic field and light forces. Since the capacity of cooling in this trap is limited, a conservative trap needs to be employed in which further cooling by evaporation can be performed. In our setup we transfer the atoms into a magnetic trap for this purpose. Since inelastic collisions with the background atoms limit the efficiency of the evaporative cooling procedure, the atom cloud is additionally transferred to another part of the vacuum chamber with lower pressure. In our experiment, this transfer is realized using magnetic forces.

One of the central achievements of the described work is the creation of a dual-species magneto-optical trap with large atom numbers as well as a magnetic transport of the cold atoms over a large distance. The large number of atoms at these stages of the gas preparation procedure is of interest not only because it allows the anticipation of the losses induced by the subsequent evaporative cooling procedure, but also because it allows making the evaporation procedure more efficient. Besides, the Fermi temperatures of the gas are larger for larger atom numbers and thus quantum phenomena can be observed at higher temperatures. Finally, a large atom number leads to better signal-to-noise ratios and a greater robustness in day-to-day operation.

In a dual-species MOT, the atom number is in general reduced compared to single-species MOTs due to additional interspecies collisions and to experimental constraints, such as the imperative to use the same magnetic field for both species or common optics. In other groups working with the ^{6}Li-^{40}K mixture the following system performances have been achieved: in the Munich group [89] the dual-species MOT is loaded from a Zeeman slower for ^{6}Li and a vapor for ^{40}K, resulting in $\sim 4 \times 10^7$ trapped ^{6}Li and $\sim 2 \times 10^7$ ^{40}K atoms. In the Innsbruck group [43] the dual-species MOT is loaded from a multi-species Zeeman slower and atom numbers of $\sim 10^9$ (^{6}Li) and $\sim 10^7$ (^{40}K) are achieved. In the Amsterdam group [44] two separate two-dimensional (2D) MOTs allow loading of $\sim 3 \times 10^9$ ^{6}Li and $\sim 2 \times 10^9$ ^{40}K atoms. In the MIT group [42] the dual-species MOT is loaded from two separate Zeeman slowers and atom numbers of

$\sim 10^9$ (^{6}Li) and $\sim 5 \times 10^7$ (^{40}K) are achieved. In our setup, the dual-species MOT is loaded from a Zeeman slower for ^{6}Li and a 2D-MOT for ^{40}K. It simultaneously contains 5.2×10^9 ^{6}Li and 8.0×10^9 ^{40}K atoms, which represents a substantial improvement in the performance.

For our application in particular a large atom number in the ^{40}K-MOT is of interest, since we intend to sympathetically cool ^{6}Li with ^{40}K, where ^{40}K will be prepared and cooled in two different spin states in a magnetic trap. This approach has been implemented by Tiecke and coworkers [92] and proved to be an efficient cooling method, as it can be realized in a magnetic trap, which provides a large trapping volume and steep confinement. In this cooling process mostly ^{40}K atoms will be lost.

The constructed experimental apparatus is described in **chapter 2**. We in particular detail the description of the implemented atom sources (the Zeeman slower for ^{6}Li and the two-dimensional magneto-optical trap for ^{40}K), as they yield high fluxes of cold atoms. Furthermore we present the details on the optimum operation of the dual-species MOT. Besides, we present the implementation of the magnetic transport system, with which the atom clouds are transferred to an ultra-high vacuum environment for evaporative cooling to quantum degeneracy. The evaporative cooling will be performed in an optically plugged magnetic trap whose implementation is in progress.

The characterization of the constructed apparatus is presented in **chapter 3**. The performances of the atomic beam sources, the magneto-optical trap and the magnetic transport are determined and their dependence on a series of parameters are presented. Furthermore, we present a study of light-induced interspecies collisions in the dual-species MOT and describe the applied strategy to minimize these collisions. Parts of the work presented in chapters 2 and 3 have been published in Ref. [122] (see Appendix D).

The central scientific result of this book is the first creation of excited heteronuclear ^{6}Li^{40}K* molecules by single-photon photoassociation. The experiment and its analysis are presented in **chapter 4**. We performed photoassociation trap loss spectroscopy in the dual-species magneto-optical trap, probing the most weakly bound rovibrational states, and we identified the observed resonances. The long-range dispersion coefficients of the different excited molecular potentials and the rotational constants are derived from the spectra. We find large molecule formation rates of up to $\sim 3.5 \times 10^7 \text{s}^{-1}$, which are shown to be comparable to those for homonuclear ^{40}K$_2^*$. Using a theoretical model we infer decay rates to the deeply bound electronic ground-state vibrational level $X^1\Sigma^+(v' = 3)$ of $\sim 5 \times 10^4 \text{s}^{-1}$. Our results pave the way for the production of ultracold bosonic ground-state ^{6}Li^{40}K molecules which exhibit a large intrinsic permanent electric dipole moment. In this chapter, we also present the novel results obtained from photoassociation spectroscopy of ^{40}K$_2^*$ and compare them to prior studies with ^{39}K$_2^*$. Parts of the work presented in chapter 4 have been published in Ref. [123] (see Appendix D).

Another scientific result of this book is the theoretical study of the motion of quantum particles inside rapidly oscillating potentials, which is described in **chapter 5**. We present and investigate a novel method to manipulate the mean motion of quantum particles in such potentials, which consists in instantaneously changing the potential's phase. This method could be applied to polar ^{6}Li^{40}K molecules. Analytical calculations show that this method allows for a significant manipulation of the particle's mean motion in a controlled way. The work presented in this chapter has been published in Refs. [124, 125] (see Appendix D). The described method has already been implemented experimentally by another research group using a Bose-Einstein condensate trapped in a time-averaged orbiting potential [126].

Chapter 2

Construction of the experimental apparatus

The objective of the current chapter is to give a basic description of the experimental Fermi mixture apparatus, whose design and construction represented the most time-consuming part of the work described in this book. We started the construction process with a completely empty laboratory and it took three and a half years to build up the machine from scratch.

When developing a new machine for ultracold atom experiments a careful analysis of the possible experimental strategies is very important and decisive for its future performance. Many techniques to produce ultracold atoms have been developed over the past years, mostly for single species. For a machine that is designated to produce mixtures of two different species, combining certain techniques for the two different species can lead to a significant simplification of the system. However, simplifications very often have to be paid with a reduction of the achievable atom number. The main goal of our Fermix machine is to produce degenerate mixtures of ultracold fermions with large atom numbers and with a reasonable repetition rate. We therefore implemented for each atomic species the technique which yields the largest amount of atoms and only then combined certain techniques for simplification if those do not significantly restrict the atom numbers.

2.1 Design considerations

In order to create a quantum degenerate ultracold gas, the atoms must be cooled and compressed in a trap until the deBroglie wavelength is of the order of the spacing between atoms. In order to achieve this the atoms are typically precooled by laser cooling and the final cooling is realized by forced evaporative cooling. Most of today's ultracold atom experiments employ a magneto-optical trap (MOT) as a precooling stage. We intended to do so as well and the first decision we had to take was how to load the MOT. In general, a MOT can be either loaded from a beam (continuous or pulsed) of slow atoms, or from a background vapor. The loading from a background vapor is very convenient, however, it is typically accompanied by short trap lifetimes, since the vapor pressure needs to be high in order to obtain sufficient loading rates. One approach to overcome this limitation is pulsed loading using an alkali getter dispenser [127] or ultraviolet light-induced absorption [128, 129]. However, the typical obtainable lifetimes and loading rates can still not compete with those which can be achieved using a beam of slow atoms. In this scheme the MOT can be loaded in an ultra-high vacuum environment and large beam fluxes can be obtained. Therefore we decided to implement this MOT loading scheme. We decided further to implement a separate source for each species. We implemented a Zeeman slower for ^{6}Li and a 2D-MOT for ^{40}K.

The repetition rates of the experiment and the trap lifetimes can be further increased by transporting the cloud into another chamber with even lower pressure. Since the experimental procedure is continued in another chamber, the MOT can already be reloaded before the previous procedure is over. The transport of a cold atomic cloud can be done either using optical dipole forces [130] or magnetic forces [131]. The trapping potentials induced by optical forces have small volumes compared to the magnetic case. For an efficient optical transport the atom cloud would therefore first need to be significantly compressed by an evaporative precooling stage prior to the transport. Due to the restricted vacuum quality in the chamber of the MOT this leads to significant atom loss. A transport of the atoms based on magnetic forces makes it possible to nearly immediately move the atoms into the better vacuum environment of the final cell, allowing one to perform the time-consuming evaporative cooling procedure there. Based on these considerations we decided to implement a magnetic transport. Further we decided to perform the evaporative cooling procedure in an optically plugged magnetic quadrupole trap, since that allows for an efficient cooling procedure due to the large trapping volume and the steep confinement of the trap. Furthermore, the trap gives a large optical access to the atoms in the final cell.

Currently, four more groups are working with the fermionic mixture ^{6}Li-^{40}K. The Munich group [89] (now Singapore) has a triple-species machine producing a mixture of ^{6}Li, ^{40}K and ^{87}Rb. The multi-species MOT is loaded from a Zeeman slower for ^{6}Li and a vapor for ^{40}K and ^{87}Rb. The mixture is transported magnetically to a UHV chamber and the cloud is transferred into an Ioffe-Pritchard trap, in which it is evaporatively cooled to quantum degeneracy. The cooling is based on sympathetic cooling of the species ^{6}Li and ^{40}K with the evaporatively cooled species ^{87}Rb. The Innsbruck group [43] loads the ^{6}Li-^{40}K dual-species MOT from a multi-species Zeeman slower, which also allows adding a third fermionic species, ^{87}Sr, to the system. The ^{6}Li-^{40}K cloud is directly transferred into an optical dipole trap and evaporatively cooled in the MOT chamber to quantum degeneracy. The cooling is realized by evaporatively cooling ^{6}Li in two different spin states and sympathetically cooling ^{40}K along. The group in Amsterdam [44] loads the ^{6}Li-^{40}K dual-species MOT from two separate 2D-MOTs. The cloud is transferred into an optically plugged magnetic quadrupole trap, in which it is evaporatively cooled inside the MOT chamber. The cooling is realized by evaporatively cooling ^{40}K in three different spin states and sympathetically cooling ^{6}Li along. It is then transferred into an optical dipole trap and transported into a UHV environment by an optical transport. The group at MIT [42] has a two-species-two-isotope machine producing a mixture of ^{6}Li, ^{40}K and ^{41}K. The dual-species MOT is loaded from two independent Zeeman slowers. It is transferred into an optically plugged magnetic quadrupole trap, where it is evaporatively cooled to quantum degeneracy inside the MOT chamber (whose vacuum quality is enhanced by a coating with a thermally activated Titanium-Zirconium-Vanadium alloy which acts as an efficient non-evaporable getter). The cooling is based on sympathetic cooling of the ^{6}Li and ^{40}K with the evaporatively cooled ^{41}K.

2.2 Vacuum manifold

2.2.1 Setup

A three-dimensional view of our vacuum system is shown in Fig. 2.1. It consists of two atom trap chambers and three flux regions. The first chamber is a central octagonal chamber where the ^{6}Li-^{40}K dual-species MOT is prepared. The second chamber is a glass science cell, in which we will evaporatively cool the mixture to quantum degeneracy.

The three flux regions are all connected to the octagonal chamber and are divided in two parts. First, the atom sources, namely a 2D-MOT for ^{40}K and a Zeeman slower for ^{6}Li. Second, a magnetic transport connecting the octagonal chamber to the final science cell. This magnetic transport consists of a spatially fixed assembly of magnetic coils which creates a moving trapping potential of constant shape by applying time-varying currents [131].

The octagonal chamber was designed by our group and manufactured by the company Caburn-MDC Europe (for engineering drawings see Appendix C.1). It can be

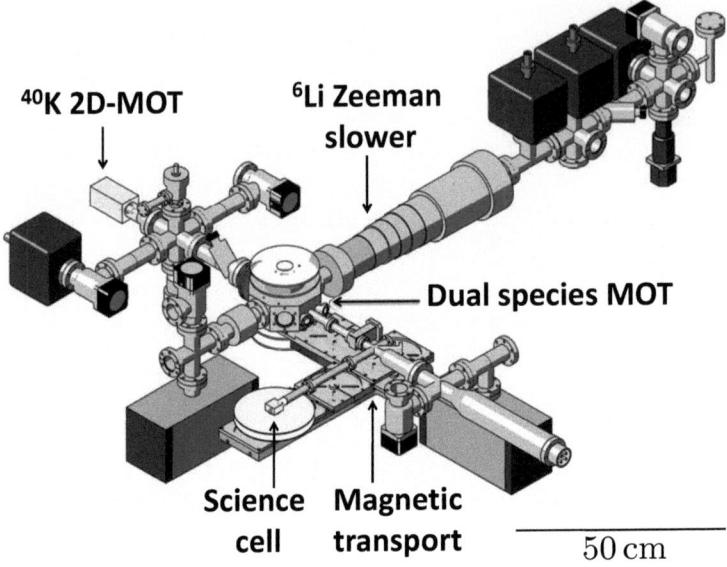

Figure 2.1: Schematics of the vacuum assembly. The dual-species MOT is loaded from a 2D-MOT for ^{40}K and a Zeeman slower for ^{6}Li. A magnetic transport allows the transfer of the cloud to a UHV science cell with large optical access.

isolated from the source regions and the science cell by all-metal ultra-high vacuum (UHV) valves, which allow for separate baking and trouble-shooting. The 2D-MOT and the Zeeman slower region are pumped by one and three 20 l/s ion pumps, respectively. The octagonal chamber is pumped by a 40 l/s ion pump and the science chamber by a 40 l/s ion pump and an occasionally launched titanium sublimation pump. Differential pumping tubes connect the source regions to the octagonal chamber in order to create a high vacuum environment in the octagonal cell. In a similar way, the science chamber is connected to the octagonal chamber via a combination of standard CF16- and homemade vacuum tubes of 1 cm diameter to further increase the vacuum quality. These tubes were designed to allow most of the magnetically transported atoms to pass through and to limit the conductance enough to have an adequate pressure in the science cell.

The science cell was designed by our group and manufactured by Hellma GmbH (for engineering drawings see Appendix C.2). A photograph of the cell is shown in Fig. 2.2. It is made out of Vycor, a synthetic fused silicon dioxide. Its glass-to-metal junction, which allows the attachment of the glass cell to the rest of the vacuum

2.2. Vacuum manifold

chamber, is non-magnetic. The walls of the glass cell have a uniform thickness of 4 mm with a manufacturing precision of ±0.01 mm. The inner cell dimensions are 23 mm×23 mm×10 mm. They have been chosen to be of that small size in order to provide large mechanical and optical access and to obtain small wall deformations due to the pressure difference. According to the manufacturer's specifications, at a pressure difference of ∼ 1 bar the larger surfaced cell walls deform by ∼ 770 nm toward the inside and the smaller surfaced walls by ∼ 170 nm toward the outside of the cell (this information will be important for the implementation of a high-resolution imaging system). The small cell dimensions permit to place magnetic coils at a close distance from the atoms, with which a steep magnetic confinement can be achieved, desirable for efficient evaporative cooling. Furthermore, an objective with a large numerical aperture can be installed for high-resolution imaging.

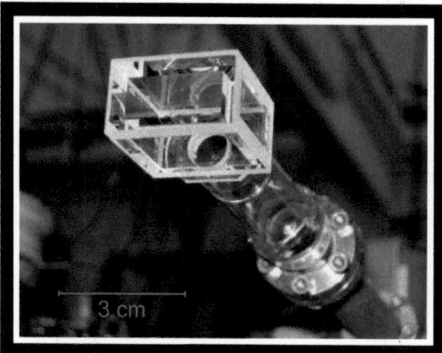

Figure 2.2: Photograph of the ulra-high vacuum science cell in which the ultracold atomic mixture is prepared and investigated. It was designed to have a small size in order to provide large mechanical and optical access and to reduce the wall deformations due to the pressure difference.

2.2.2 Assembly, pump down and bake out

Obtaining UHV pressures requires a careful assembly of the vacuum components, and a baking of the chamber at high temperatures under vacuum. We describe in the following the crucial steps of the assembly and the subsequent baking process. Prior to assembly it has to be made sure that all vacuum components are clean. Dirty components are cleaned in an ultrasonic cleaner. During the assembly powder-free latex gloves are worn and changed frequently to avoid contamination. The system is mounted loosely to the optical table to avoid stress on the system's gaskets during assembly and the subsequent baking. A turbo molecular pump and a residual gas analyzer (RGA) are connected to the system for the pumping procedure. When the system is sealed the turbo molecular pump is run for ~ 2 hours. If the residual gas pressure, which is in general determined by the partial pressure of hydrogen, falls below $\sim 5 \times 10^{-7}$ mbar, it can be concluded that the system has no major leaks. Otherwise a leak test with helium needs to be performed to find the leak and close it. Then the baking procedure begins.

First, the glass cells are wrapped with clean fiberglass cloth for protection. Next, thermocouples are placed on the vacuum system at critical places such as cells, valves, glass-to-metal seals and pumps in order to control their temperature during the bake out. Then the system (including the ion pumps with the magnets in place) is wrapped with several resistive heater tapes, which are powered by variable AC transformers (Variacs). In addition the system is wrapped loosely with many layers of aluminum foil in order to obtain a homogeneous temperature around the chamber. The system is then slowly brought up to the final temperature over at least 12 hours with a maximum temperature change of 1°C/min. We heated our system to temperatures between $220 - 250$°C. The ion pumps are off during the warm up to minimize their pollution. It has to be made sure that temperature gradients, which apply stress to the system, are kept low across the glass cells, glass windows and glass-to-metal junctions, which are the most fragile components. Throughout the warming up process, thermocouple readings and Variac settings are recorded for surveillance and to facilitate future bakes.

Once the system is at the desired temperature, it is baked with just the turbo pump for 2 days. At this point also the titanium sublimation pump is run several times to degas the filaments. Our titanium sublimation pump has three filaments to which a maximum current of 50 A can be applied. They are each subsequently run 2 times for one minute at 30 A, once for one minute at 40 A and once for two minutes at 40 A and finally once for 2 h at 30 A. After each titanium sublimation shot the pressure increased to $\sim 10^{-4}$ mbar, which required to wait for the pressure to fall again before launching the next shot. Then the titanium sublimation pump is kept off, the ion pumps are turned on and the system is pumped like this for several more days. We start to cool the system down only when the hydrogen pressure falls below $\sim 10^{-7}$ mbar. The partial pressures of the other gases (H_2O, N_2, O_2, CO_2, etc.) are typically two orders of magnitude less ($\sim 10^{-9}$ mbar). The system is cooled down slowly over ~ 8 h to avoid excessive temperature gradients. After the bakeout the pressure in our system, as indicated by the RGA, was of the order of 1×10^{-10} mbar. The ion pump currents can usually be used to determine the pressure. For the low final pressure, however, this does not work, since each ion pump controller has an (unknown) offset, which slightly changes over time. At room temperature the turbo pump and the RGA are valved off and shut down, which improves the ultimate pressure and minimizes vibrations. The final pressure in the different parts of the system is most precisely measured in terms of the inverse lifetime of trapped atoms.

2.3 Laser systems

Our experimental apparatus requires separate laser systems and optics for the two different atomic transition wavelengths 671 nm (Li) and 767 nm (K). The laser systems provide several beams with different frequencies and intensities for slowing, trapping and probing each atomic species. A sketch of the energy levels of the atomic species and the frequencies of interest are shown in Fig. 2.3. The laser systems of the two atomic species are set up on separate optical tables and the generated laser beams are transferred to the main experimental table using optical fibers.

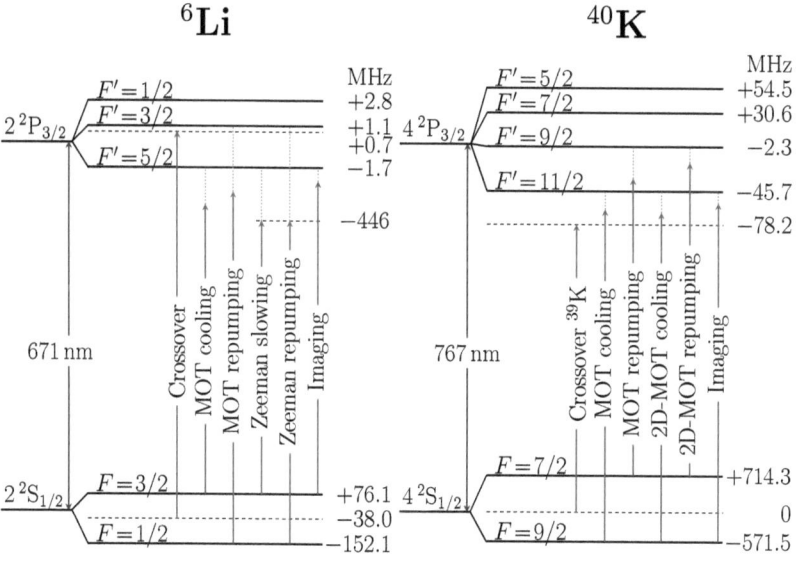

Figure 2.3: Level diagrams for the ^{6}Li and ^{40}K D$_2$-lines with their respective hyperfine structures, showing the frequencies required for the dual-species MOT operation. The diode lasers are locked to the indicated saturated absorption crossover signals $2S_{1/2}(F=1/2, F=3/2) \rightarrow 2P_{3/2}$ of ^{6}Li and $4S_{1/2}(F=1, F=2) \rightarrow 4P_{3/2}$ of ^{39}K.

2.3.1 Optics

A simplified scheme of the laser systems is shown in Fig. 2.4. Each one consists of a single low output-power frequency-stabilized "master" diode laser (DL) and three tapered amplifiers (TAs) used for light amplification. Due to the small hyperfine splittings of both ^{6}Li and ^{40}K, the required frequencies of the various laser beams are conveniently shifted and independently controlled by acousto-optical modulators (AOMs).

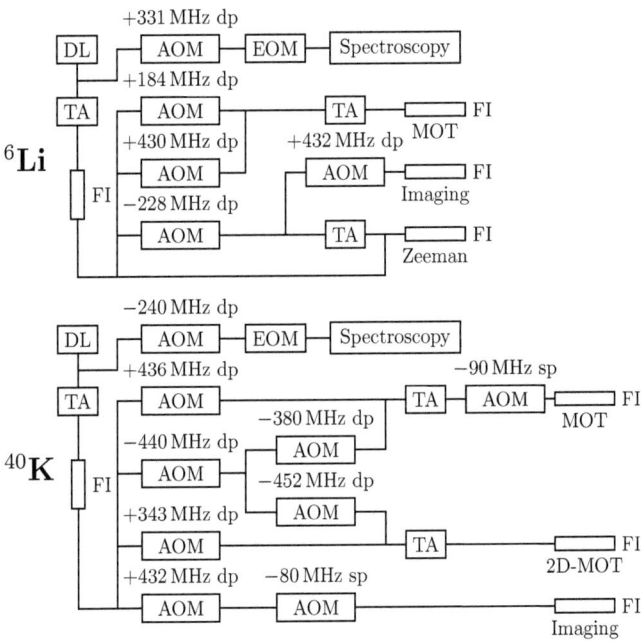

Figure 2.4: Laser systems for ^{6}Li and ^{40}K. The frequencies and amplitudes of the various beams are controlled by AOMs in single pass (sp) or double pass (dp) configuration. The EOMs are used to phase modulate a part of the beam for the diode laser's frequency stabilization. Single mode polarization maintaining fibers (FI) are used for beam shaping and spatial filtering. The indicated AOM frequencies allow the generation of the beam frequencies required for the dual-species MOT (see Fig. 2.3).

In both laser systems the frequency stabilized master laser beam is immediately amplified by a first TA and subsequently injected into a single-mode polarization maintaining optical fiber (FI) for beam shaping and spatial filtering (see Fig. 2.4). The output beam of the optical fiber is split by a series of polarizing beam splitters into several beams whose frequencies and intensities are independently shifted and con-

2.3. Laser systems

trolled with AOMs in single or double pass configuration. The various beams are then recombined with a pair of polarizing beam splitters to linearly polarized bichromatic beams consisting of one cooling and one repumping frequency. Those are then either directly injected into a fiber or into another TA for further amplification. The fibers finally transfer the beams to the main experimental table. A photograph of the optical table for lithium is shown in Fig. 2.5.

Figure 2.5: Photograph of the optical table for lithium, showing the optical components required to implement the scheme of Fig. 2.4. In the photograph the laser beams have been made visible by cigarette smoke.

The injection of a bichromatic beam into a TA, whose gain-medium is non-linear, is accompanied with the creation of sidebands [132]. The sideband creation is due to parametric amplification of the gain medium by the beating between the two injected frequencies. In general, sidebands represent a loss of the power available in the injected frequencies and can excite unwanted transitions. The power of the sidebands depends on the difference in power and frequency of the injected beams—it is larger for smaller differences of both quantities. In our case, where the two injected beam components have significantly different powers and frequencies (differing by ~ 228 MHz for ^{6}Li and by ~ 1286 MHz for ^{40}K), the power losses to the sidebands are small: for a power ratio of $1/5$ of the injected components, the resulting total power of the sidebands is $\sim 5\%$ for ^{6}Li and $\sim 0.2\%$ for ^{40}K. The frequency of the sidebands depend on the difference in frequency of the injected beams. Unwanted transitions are excited when the sidebands are near-resonant with an atomic transition. For the bichromatic beam pairs used for the MOTs and the 2D-MOT, no unwanted transitions are excited. For the bichromatic beam used for the Zeeman slower, however, a sideband would be near-resonant with the atoms trapped in the ^{6}Li-MOT, as the Zeeman beams are detuned by approximately an integer multiple of 228 MHz. For this beam the injection of both frequency components into the same TA was thus avoided (see Fig. 2.4).

Acoustically isolated homemade mechanical shutters are placed in front of each fiber on the optical tables allowing us to switch off the laser beams when required. The shutters consist of a low-cost solenoid-driven mechanical switch (Tyco Electronics, ref. T90N1D12-12) and a razor blade attached to it via a small rigid lever arm. These shutters typically have a closing time of $\sim 100\,\mu$s when placed in the focus of a laser beam and a time delay of the order of 3 ms with a timing jitter of ± 1 ms.

2.3.2 Diode lasers

The diode lasers are homemade tunable external cavity diode lasers in Littrow configuration. The laser diode for Li (Mitsubishi, ref. ML101J27) is of low cost due to its mass production for the DVD industry. Its central free running output wavelength at room temperature is 660 nm which can be shifted into the range of 671 nm by heating the diode to $\sim 80°$C. In external cavity configuration its output power is 40 mW at a driving current of 150 mA. Under these conditions the laser diode reaches a typical lifetime of 6 months. It can be mode hop-free tuned over a range of 5 GHz.

The laser diode for K is an anti-reflection coated Ridge-Waveguide Laser (Eagleyard, ref. EYP-RWE-0790-0400-0750-SOT03-0000), whose central free running output wavelength at room temperature corresponds to the desired wavelength. In external cavity configuration its output power is 35 mW at 90 mA and it can be tuned between 750 nm and 790 nm. The mode hop-free tuning range extends over 10 GHz. The diode has a typical lifetime of one year.

The frequency of each diode laser is stabilized via saturated absorption spectroscopy for which a small part of the DL's output is used (see Fig. 2.1). A 20 MHz electro-optical modulator (EOM) is employed to modulate the phase of the spectroscopy laser beam yielding the derivative of the absorption signal through a lock-in detection. The resulting error signal is transferred to both the diode's current (via a high frequency bias-tee), and, via a PID-controller, to a piezo that adjusts the external cavity's length with a 4 kHz bandwidth. An AOM is used to offset the frequency of the diode laser with respect to the absorption line used for locking. It allows for fine adjustments of the frequency while the laser is locked.

The Li diode laser frequency is shifted by -331 MHz from the ^{6}Li $2S_{1/2}(F = 1/2, F = 3/2) \rightarrow 2P_{3/2}$ crossover signal and the K diode laser frequency is shifted

2.3. Laser systems

by +240 MHz from the conveniently located $4S_{1/2}(F = 1, F = 2) \to 4P_{3/2}$ crossover signal of ^{39}K. Note that the small excited-state hyperfine structures of both ^{6}Li and ^{39}K are unresolved in the spectroscopy. The AOMs are driven by homemade voltage-controlled oscillators, whose outputs are amplified using rf-amplifiers (Minicircuits, ref. ZHL-1-2W).

2.3.3 Saturated absorption spectroscopy

The saturated absorption spectroscopy for lithium is realized in a 50 cm-long heat pipe, in which a natural Li sample (with the isotopic abundances ^{7}Li: 92%, ^{6}Li: 8%) is heated to $\sim 350°$C to create a sufficiently high vapor pressure for absorption. The heat pipe consists of a standard CF40 tube which is closed at each end with a window. The Li-sample is placed at its center. The tube is heated with a pair of thermocoax cables which are wound around the tube in parallel with opposite current directions in order to prevent magnetic fields to build up. Condensation of lithium atoms on the cell windows needs to be inhibited as Li chemically reacts with glass. This is achieved by adding an argon buffer gas at ~ 0.1 mbar pressure, as Ar-Li collisions prevent Li to reach the cell windows in ballistic flight. The optimum argon pressure was chosen such that it provides enough collisions, but does not substantially collision-broaden the absorption spectrum. Water cooling of the metallic parts close to the windows leads to condensation of the diffusing lithium atoms before those can reach the windows. To avoid that lithium slowly migrates to the colder surfaces, the inside of the tube is covered with a thin stainless steel mesh (Alfa Aesar, ref. 013477), which induces capillary forces acting on the condensed atoms. Since the surface tension of liquid lithium decreases with increasing temperature [133], the capillary forces cause the atoms to move back to the hotter surfaces.

The saturated absorption spectroscopy for potassium is realized in a cylindrical glass vapor cell of 5 cm length, in which a natural K-sample (with the isotopic abundances ^{39}K: 93.36%, ^{40}K: 0.012%, ^{41}K: 6.73%) is heated to $\sim 45°$C. Here, a small non-heated appendix of the cell serves as a cold point to prevent condensation of K atoms on the surfaces crossed by the laser beam.

2.3.4 Tapered amplifiers

The tapered amplifiers are commercial semiconductor chips which are mounted on homemade supports (for engineering drawings, see Appendices C.3 and C.4). We developed compact support designs with nearly no adjustable parts. The support designs allow for an easy installation process, which does not require any gluing or the help of micrometric translation stages for the alignment of the collimation optics, as that can be accomplished by free hand. Furthermore, the design minimizes the heat capacity of the support and the produced temperature gradients, allowing for a quick temperature stabilization that makes the TAs quickly operational after switch-on. The temperature stabilization is accomplished using a Peltier element (Roithner Lasertechnik GmbH, ref. TEC1-12705T125) connected to a PID control circuit. The heat of the chip is dissipated via an aluminum base plate which is economically cooled by air rather than running water (the base plate reaches a maximum temperature of $\sim 28°C$ for diode currents of 2A).

The commercial TA chips are sold on small heat sinks which have different dimensions for the two different wavelengths. We thus had to design slightly different supports for the Li- and K-TAs, which are both schematically shown in Fig. 2.6.

For lithium the semiconductor chip (Toptica, ref. TA-670-0500-5) is delivered on a heat dissipation mount of type "I" (see Fig. C.9 of the appendix). It is placed between two axially aligned cylindrical lens tubes (CL1 and CL2 in Fig. 2.6 (a)), each of which containing an aspheric collimation lens of focal length 4.5 mm (Thorlabs, ref. C230TME-B). The support of the tubes and the chip are precisely machined such that the chip's output beam falls on the center of the respective collimation lens (CL2 in Fig. 2.6 (a)). The tubes are supported by cylindrically holed tightenable hinges in which they can move only longitudinally, along the direction of the amplified laser beam. This restriction of the tube's motion facilitates the alignment of the collimation lenses. The support design does not allow for a transverse alignment of the collimation lenses. Since this alignment is not very critical for the performance of the TA, we found it needless to allow this degree of freedom and relied on precise machining (possible imperfections could be compensated utilizing the mechanical play of the large attachment screw holes of the commercial heat sinks of the chips). When tightened by a screw, the hinges fix the position of the tubes. Since the tightening applies a force perpendicular to the longitudinal direction, it does not move the tubes along this (critical) direction. They might only move slightly along the transverse direction, which does not affect the final performance of the TA.

For potassium, the semiconductor chip (Eagleyard, ref. EYP-TPA-0765-01500-3006-CMT03-0000) is delivered on a heat dissipation mount of type "C" (see Fig. C.3 of the appendix). Placing this mount between two hinges as for the case of lithium is less convenient since the heat dissipation mount has to be attached by a screw in the longitudinal direction which requires access from one side. Therefore one hinge is replaced by a rail which guides a parallelepipedically formed mount for the second (output) collimation lens (CL4 in Fig. 2.6 (b)). The motion of this mount is also fixed

2.3. Laser systems

by tightening a screw applying forces perpendicular to the rail direction, which does not move the collimation lens along the critical longitudinal direction. For all our TAs, the positioning of the collimation lenses never had to be adjusted again once they were aligned.

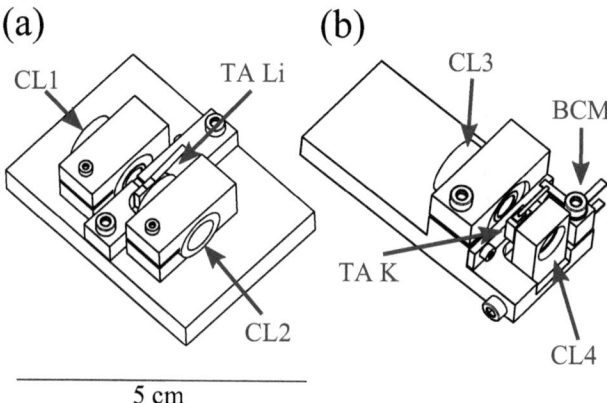

Figure 2.6: Sketch of the tapered amplifier supports for (a) Li and (b) K. In the figure, TA Li and TA K refer to the respective tapered amplifier chips, CL1, CL2, CL3 and CL4 to the (only longitudinally adjustable) collimation lens supports and BCM to the isolated mount for the blade connectors used to power the chip for K. The supports for the output collimation lenses are CL2 and CL4.

The commercial heat dissipation mount of the potassium chip is inconvenient for a simple powering of the chip. The very fragile gold wire, which has to be connected to the negative source of the current supply, has to be protected by a mechanical support before being connected to a cable. Therefore we soldered it to a blade connector that is fixed by an isolated plastic mount (BCM Fig. 2.6 (b)) and which is connected to the current supply. To avoid an overheating of the chip during the soldering process we permanently cooled the gold wire by blowing cold dry air from a spray can on it.

The output beams of the TA chips are astigmatic and thus require additional collimation. The choice of the collimation optics needs to be adapted to the specifications of the subsequent optical fiber, which in our case requests a collimated circular Gaussian beam of 2.2 mm $1/e^2$-diameter for optimum coupling efficiency. The mode-matching was found optimum for a pair of lenses consisting of one spherical lens (with f=15 cm for Li and f=4 cm for K) and a cylindrical lens (with f=8 cm for Li and f=2.54 cm for K), which are placed outside the TA's housing. The cylindrical lenses are supported by rotatable mounts, in order to facilitate the mode-matching into the fibers. For all our TAs we achieve fiber-coupling efficiencies larger than 50% (Li) and 60% (K).

When injected with 20 mW, the Li-TAs yield an output power of 500 mW at 1 A driving current and the K-TAs yield an output power of 1500 mW at 2.5 A driving current. In order to increase the lifetime of the chips, we limit the driving currents to smaller values and we switch the chips on only for periods of experimentation. When switched on, the TAs quickly reach a stable functioning (usually within 10 min) due to the compactness of the mechanical support, which allows for a quick temperature stabilization.

2.4 ^{6}Li Zeeman slower

2.4.1 Principle of Zeeman-tuned slowing

Zeeman-tuned slowing represents one of the earliest and most widely used techniques to slow down atoms from an oven [134]. Many textbooks nowadays treat this cooling technique and we refer the reader to the literature for an introduction [135]. A Zeeman slower longitudinally decelerates an atomic beam using the radiative force of a counter-propagating resonant laser beam. The Doppler effect accumulated during the deceleration is compensated by the Zeeman effect, induced by an inhomogeneous magnetic field, which maintains the atoms close to resonance and provides a continuous deceleration.

Two types of Zeeman slowers are commonly used: the positive-field and the sign-changing field ("spin-flip") Zeeman slower [135]. We have implemented a spin-flip Zeeman slower since it brings about several advantages. First, a smaller maximum absolute value of the magnetic field is required. Second, the Zeeman laser beam is non-resonant with the atoms exiting the slower and thus does not push them back into the slower, neither does it perturb the atoms trapped in the ^{6}Li-MOT. Finally, less coil windings are required close to the MOT, allowing for a better optical access. However, the spin-flip Zeeman slower requires repumping light in the region where the magnetic field changes sign and thus makes the optics system slightly more complicated.

The Zeeman slower consists of two distinct parts: the oven, which creates an atomic beam of thermal atoms, and an assembly of magnetic field coils. The construction of both parts is described in the following.

2.4.2 Oven

The design of the oven is based on the one described in the thesis of Florian Schreck [136]. In the oven a nearly pure ^{6}Li sample (5 g) is heated to $\sim 500°C$ and an atomic beam is

2.4. ^{6}Li Zeeman slower

extracted through a collimation tube. The oven consists of a vertical reservoir tube (diameter: 16 mm, length: 180 mm) and a horizontal collimation tube (diameter: 6 mm, length: 80 mm), which is attached to it (see rightmost component in Fig. 2.1). The upper end of the reservoir tube and the free end of the collimation tube are connected to CF40-flanges. The flange of the reservoir tube is sealed and allows connecting a vacuum pump for baking purposes. The flange of the collimation tube connects the oven to the rest of the vacuum chamber. All parts of the oven are made of stainless steel of type 302L and connected using nickel gaskets instead of copper gaskets as they stand higher temperatures and react less with lithium. The heating of the oven is realized with two high power heating elements (Thermocoax, ref. SEI 10/50-25/2xCM 10), wound around both, the reservoir and the collimation tube.

The temperature of the oven needs to be stabilized precisely, since the atomic flux critically depends on the temperature. This is accomplished by an active stabilization circuit and an isolation with glass wool and aluminum foil. Along the collimation tube a temperature gradient is maintained in order to recycle lithium atoms sticking to the inner tube walls through capillary action, as explained above. In order to amplify the effect of capillary action, a thin stainless steel mesh with a wire diameter of 0.13 mm (Alfa Aesar, ref. 013477) is placed inside the tube. This wire decreases the effective diameter of the collimation tube to ~ 5 mm. For the operating temperature of 500°C, the vapor pressure of lithium in the oven amounts to 4×10^{-3} mbar.

A computer controlled mechanical shutter (Danaher Motion, ref. BRM-275-03) in front of the oven allows blocking of the atomic beam during experiments or when the ^{6}Li-MOT is not in operation. Collisions between the atomic beam and the trapped atoms can thus be inhibited. Furthermore, it allows preventing the lithium atoms from covering and destroying the entrance window of the Zeeman slower beam, which is placed opposite to the oven (without having to switch off the oven). Finally, the shutter facilitates the initial alignment of the Zeeman slower beam, as that is made visible by reflection on the blocking surface.

The oven is pumped through the collimation tube with a 20 l/s ion pump and isolated from the main chamber via three differential pumping stages and the tube of the Zeeman slower. The pressure drop created by the tubes can be calculated as follows. The conductance C of a tube of circular cross section in the molecular flow regime (*i.e.* where the mean free path of a particle is greater than the tube diameter) is given by [137]

$$C = 2.6 \times 10^{-4} \, \bar{v} \, \frac{D^3}{l} \quad \text{L/s}, \tag{2.1}$$

where D and l are, respectively, the diameter and length of the tube in units of cm and $\bar{v}$ is the average molecular velocity in units of cm/s. For air at 20°C, the conductance is $C = 12 D^3/l$ in units of L/s. The net pumping speed S of a pump across the tube is given by $S = (1/S_\mathrm{p} + 1/C)^{-1}$, where S_p is the nominal pumping speed of the pump and the pressure drop across the tube is given by $P_1/P_2 = S_\mathrm{p}/S$, with P_1, P_2 denoting the pressure on the respective side of the tube. The net pumping speed of the oven's collimation tube thus has the value $\sim 0.19\,\mathrm{l/s}$ resulting in a pressure drop of a factor ~ 100. The second and third differential pumping tubes both have a length of 100 mm and a diameter of 5 mm and 10 mm, respectively. A 20 l/s ion pump is placed after each tube. In total a pressure drop of a factor of $\sim 2.5 \times 10^6$ between the oven and the main chamber is obtained.

The assembly of the oven is a three-step procedure. First, the metallic parts of the oven are pre-baked at $\sim 600°\mathrm{C}$ during 48 h. Then, the oven is filled with the lithium sample under air atmosphere and baked again at $\sim 600°\mathrm{C}$ during 12 h in order to eliminate the impurities in the lithium sample (mostly LiH). Typically 50% of the sample is lost during this procedure. Then, the oven is connected to the rest of the vacuum chamber under an argon atmosphere, since argon does not react with lithium. Since argon damages ion pumps, the vaccum chamber is first pumped by a turbo molecular pump during 12 h before the ion pumps are finally launched and the oven is operational.

2.4.3 Coil assembly

The Zeeman slower coils create an inhomogeneous field along the flight direction of the atoms. The coils are mounted on a 65 cm long standard CF40 tube placed between the oven and the MOT chamber. A sketch of the coil assembly and the generated axial magnetic field profile are shown in Fig. 2.7. The coil assembly extends over $L = 55\,\mathrm{cm}$ and is separated from the position of the MOT by 16 cm. The coils are connected in series and were designed such that the desired magnetic field profile is generated for a moderate driving current of 12 A. The axial magnetic field of the slower is measured to be 570 G at the entrance and -220 G at the exit.

The magnetic field of the Zeeman slower is non-zero at the position of the MOT and hence compensated by a coil placed opposite to the slower coils at a distance of 12.7 cm from the MOT (see Fig. 2.7). The compensation coil consists of 4 coil layers wound around a 10 cm long CF40 standard tube. They are powered by a separate power supply for fine adjustments. When compensated, the magnetic field has an axial gradient of 0.5 G/cm at the position of the MOT.

2.4. ⁶Li Zeeman slower

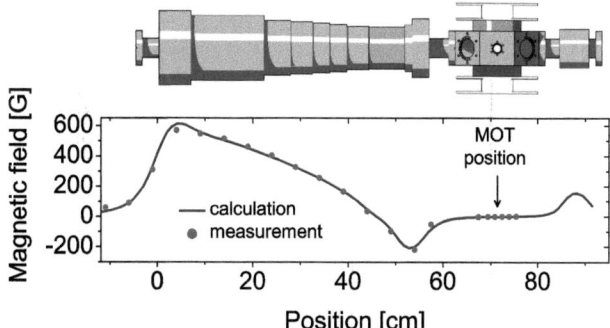

Figure 2.7: ⁶Li Zeeman slower coil assembly (top) and generated axial magnetic field profile (bottom). The thermal atoms coming from the ⁶Li-oven enter the coil assembly at the position 0, and a fraction of them is slowed down and finally captured in the ⁶Li-MOT, which is located at 71.4 cm. A compensation coil placed on the opposite side of the MOT (at 84.1 cm) ensures that the magnetic field is zero at the position of the MOT. The coil assembly extends over 55 cm.

The winding of the Zeeman slower coils is cumbersome and cannot be performed on a set-up vacuum system. Thus, the Zeeman slower needs to stand baking procedures after being wound. The cables of the Zeeman slower coils (APX France, ref. méplat cuivre émaillé CL H 1.60 × 2.50) resist temperatures of up to 200°C. One layer of a heating cable (Garnisch, ref. GGCb250-K5-19) is permanently placed underneath the magnetic field coils for the bake out procedures. To avoid heating of the vacuum parts during the Zeeman slower's operation, two layers of water coils were wound underneath the coil layers (above the heating cable). A photograph of the assembled Zeeman slower is shown in Fig. 2.8.

2.4.4 Optics

Slowing and repumping light for the Zeeman slower is derived from a bichromatic laser beam which is provided by an optical fiber originating from the laser system. It has a total power of $P_{\text{fiber}} = 50\,\text{mW}$ and its frequencies are both red detuned by $\Delta\omega_{\text{slow}} = \Delta\omega_{\text{rep}} = 75\,\Gamma$ from the $2S_{1/2}(F = 3/2) \to 2P_{3/2}(F' = 5/2)$ slowing and the $2S_{1/2}(F = 1/2) \to 2P_{3/2}(F' = 3/2)$ repumping transition (see Fig. 2.3). The intensity I_{slow} of the slowing light is 8 times bigger than the intensity I_{rep} of the repumping light. Both beam components have the same circular polarization (σ^+ at the position where the atoms enter the slower).

The detuning of the slowing light and the axial magnetic field at the entrance of

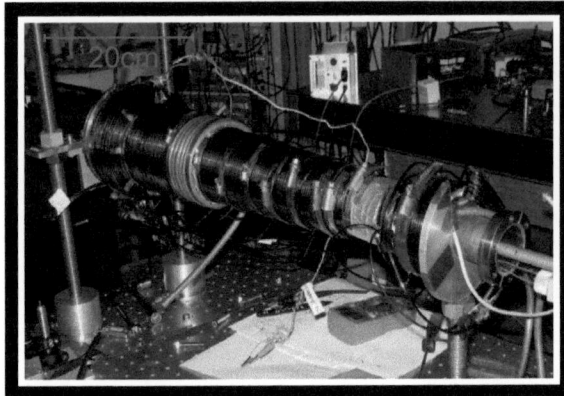

Figure 2.8: Photograph of the assembled ^{6}Li Zeeman slower before its integration into the vacuum system.

the coil assembly define the so-called capture velocity $v_{\text{cap}}^{\text{Zee}}$ of the Zeeman slower. All atoms with a velocity smaller than $v_{\text{cap}}^{\text{Zee}}$ are expected to be decelerated to the same final velocity $v_{\text{fi}}^{\text{Zee}}$, provided that they initially populate the correct internal atomic state. The resonance condition for the atoms inside the slower yields $v_{\text{cap}}^{\text{Zee}} \sim 830\,\text{m/s}$ and $v_{\text{fi}}^{\text{Zee}} \sim 90\,\text{m/s}$. The exit velocity of the slower is thus larger than the capture velocity of the ^{6}Li-MOT, which is estimated to be $\sim 50\,\text{m/s}$. However, the atoms are still decelerated significantly in the region between the slower exit and the MOT and are thus expected to be captured by the MOT. The capture velocity of the Zeeman slower is smaller than the most probable thermal speed of the atomic beam, which is given by [138] $v_p = \sqrt{2k_\text{B}T/m} = 1464\,\text{m/s}$ at $T = 500°\text{C}$, where k_B denotes the Boltzmann constant and m the mass of the ^{6}Li atoms.

The bichromatic Zeeman slower beam is expanded and focused by a lens pair. The focusing of the beam accounts for the divergence of the atomic beam and the loss of beam power due to absorption and thus yields an efficient utilization of the available laser power. In addition, it induces a small cooling effect along the transverse direction [135]. The $1/e^2$-diameter at the position of the MOT is 31 mm and the focus is at a distance of 120 cm from the MOT, 10 cm behind the oven.

The optimized values of the essential parameters used for the Zeeman slower are displayed in Tab. 2.1. With these parameters a ^{6}Li-MOT capture rate of $\sim 1.2 \times 10^9$ atoms/s is obtained. The capture rate was deduced from the measurement of the trapped atom number after a very short loading of the MOT ($\sim$250 ms), for which atom losses can still be neglected.

2.5. ^{40}K 2D-MOT

	^{6}Li Zeeman slower
P_{fiber} [mW]	50
$\Delta\omega_{\text{slow}}$ [Γ]	-75
$\Delta\omega_{\text{rep}}$ [Γ]	-75
$I_{\text{rep}}/I_{\text{slow}}$	1/8
B_{max} [G]	570

Table 2.1: Optimized values for the parameters of the ^{6}Li Zeeman slower, yielding a ^{6}Li-MOT capture rate of $\sim 1.2 \times 10^9$ atoms/s at an oven temperature of $\sim 500°$C. The definition of the symbols is given in the text. The natural linewidth of ^{6}Li is $\Gamma/(2\pi) = 5.87$ MHz.

The divergence of the atomic beam is an important parameter characterizing the Zeeman slower. Three factors contribute to it: first, the geometry of the oven's collimation and the subsequent differential pumping tubes, second the atom's deceleration inside the slower, and third the transverse heating due to the scattered photons during the slowing process. In order to estimate the divergence of the atomic beam, we calculate the maximum possible deflection of an atom which exits the oven with a longitudinal velocity $v_{\text{cap}}^{\text{Zee}}$. An atom with this velocity needs ~ 1.1 ms to reach the exit of the Zeeman slower and additional ~ 1.8 ms to reach the MOT. Due to the geometry of the collimation and differential pumping tubes it can have a maximum transverse velocity of $\sim$16 m/s. The change in transverse velocity due to the heating is calculated to be ~ 2.5 m/s [139] and is thus negligible with respect to the maximum transverse velocity determined by the tube geometry. The final transverse displacement of the atom with respect to the beam axis at the position of the ^{6}Li-MOT would thus be ~ 5 cm. Therefore ^{6}Li-MOT beams with a large diameter are required.

2.5 ^{40}K 2D-MOT

2D-MOTs have been widely used over the past years to produce high flux beams of cold atoms [140, 141, 142, 143, 144, 43]. In some cases they offer advantages over the more common Zeeman slowers. Even though Zeeman slowers can produce higher fluxes and are more robust, they have the following disadvantages. They produce unwanted magnetic fields close to the MOT which need to be compensated by additional fields, they require a substantial design and construction effort and are space consuming. Zeeman slowers need to be operated at higher temperatures than 2D-MOTs and the material consumption can be high. In the case of the rare isotope ^{40}K, this drawback is major: no pure source of ^{40}K exists and enriched ^{40}K samples are very expensive (4000 Euros for 100 mg of a 4% enriched sample). Therefore a ^{40}K Zeeman slower would be very costly. A 2D-MOT can be operated at lower pressures and is thus more economic. In addition it allows separating ^{40}K from the more abundant ^{39}K, since it produces an atomic beam which nearly only contains the slowed atoms (*i.e.* no significant thermal background). These considerations motivated us to implement a 2D-MOT for ^{40}K.

2.5.1 Principle of a 2D-MOT

In a 2D-MOT, an atomic vapor is cooled and confined transversally and out-coupled longitudinally through an aperture tube. The role of the aperture tube is two-fold. First, it isolates the 2D-MOT from the MOT chamber by differential pumping, and second, it acts as a geometric velocity filter, since only atoms with a small transverse velocity pass through. As the transverse cooling is more efficient for atoms which have a small longitudinal velocity—since those spend more time in the cooling region—most of the transversally cold atoms are also longitudinally cold. Thus, the filter indirectly filters atoms also according to their longitudinal velocity. A 2D-MOT thus produces an atomic beam which is transversally *and* longitudinally cold.

The flux of a 2D-MOT can be improved by adding a longitudinal molasses cooling to the 2D-MOT configuration [140]. Thus, the atoms spend more time in the transverse cooling region due to the additional longitudinal cooling. The longitudinal beam pair is referred to as the pushing and the retarding beam, where the pushing beam propagates in the direction of the atomic beam (see Fig. 2.9). We implemented such a configuration, making use of a 45°-angled mirror inside the vacuum chamber. This mirror has a hole at its center which creates a cylindrical dark region in the reflected retarding beam. In this region, the atoms are accelerated along the longitudinal direction by the pushing beam only, which allows an efficient out-coupling of the atomic beam.

2.5.2 Experimental setup

The vacuum chamber of the 2D-MOT consists of standard CF40 components and a parallelepipedical glass cell (dimensions 110 mm×55 mm×55 mm), which is depicted in Fig. 2.9 (for engineering drawings, see Appendix C.5). Its long axis is aligned horizontally, parallel to the differential pumping tube and the direction of the produced atomic beam. The mirror inside the vacuum chamber is a polished stainless steel mirror with an elliptical surface (diameters 3.0 cm and 4.2 cm). It is attached to the differential pumping tube inside the vacuum. It allows overlapping the two longitudinal laser beams whose powers and orientations can thus be independently controlled externally.

2.5. ^{40}K 2D-MOT

The mirror's material has a reflectivity of only 50%, but is not susceptible to chemical reactions with potassium. The differential pumping tube intercepts the mirror at its center. The tube has a diameter of 2 mm over a distance of 1.5 cm and then stepwise widens up to 10 mm over a total distance of 22 cm. The ^{40}K-MOT is located 55 cm away from the 2D-MOT center. Assuming a ballistic flight of the atoms, the geometry of the differential pumping tube defines an upper limit of the divergence of the atomic beam, which is calculated to be $\sim$35 mrad. The atomic beam thus is expected to have a diameter of $\sim$ 2 cm when it reaches the ^{40}K-MOT. The differential pumping tube has a conductance of 0.04 l/s. The generated pressure ratio between the 2D-MOT and the 3D-MOT chambers is $\sim 10^3$.

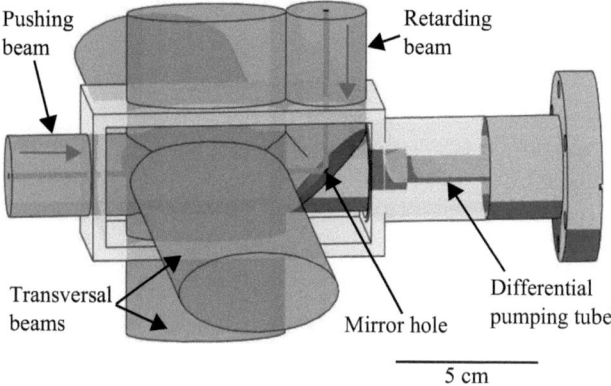

Figure 2.9: Sketch of the parallelepipedical glass cell used for the ^{40}K 2D-MOT. A mirror is placed inside the vacuum chamber to allow an independent control over the longitudinal beam pair. The mirror has a hole in its center and creates a dark cylindrical region in the reflected beams.

The potassium source is an isotopically enriched ^{40}K sample (containing 4 mg of ^{40}K, 89.5 mg of ^{39}K and 6.5 mg of ^{41}K, from Technical Glass Inc., Aurora, USA), placed at a distance of 20 cm from the glass cell. It was purchased in a small ampule which was broken under vacuum inside a modified stainless steel CF16 bellow. The small vapor pressure of potassium at room temperature (10^{-8} mbar) requires heating of the entire 2D-MOT chamber. We heat the source region to $\sim$ 100°C, all intermediate parts to $\sim$ 80°C and the glass cell to $\sim$ 45°C. The gradient in temperature ensures that the potassium migrates into the cell and remains there. The resulting K-pressure in the glass cell was measured by absorption of a low intensity probe. We found 2.3×10^{-7} mbar, which implies a partial pressure of the ^{40}K-isotope of 1×10^{-8} mbar. In contrast to lithium, the source lifetime is mainly determined by the pumping speed of the ion pump. At the measured pressure the lifetime of the source is estimated to ~ 2 years.

Four air-cooled rectangular shaped elongated racetrack coils (dimensions $160\,\text{mm}\times 60\,\text{mm}$) are placed around the glass cell to produce a 2D quadrupole field with cylindrical symmetry and a horizontal line of zero magnetic field. This racetrack coil geometry allows an independent control of the transverse position of the magnetic field zero, and minimizes finite coil fringe effects at the coil ends. The coils are controlled by four separate power supplies. For optimized operation, the transverse magnetic field gradients are $\partial_x B = \partial_y B = 11\,\text{G/cm}$.

Cooling and repumping light for the 2D-MOT is derived from a bichromatic laser beam which is provided by an optical fiber originating from the laser system. It has a total power of $P_{\text{fiber}} = 450\,\text{mW}$ and its frequencies are red detuned by $\sim 3.5\,\Gamma$ from the $4S_{1/2}(F=9/2) \to 4P_{3/2}(F'=11/2)$ cooling and by $\sim 2.5\,\Gamma$ from the $4S_{1/2}(F=7/2) \to 4P_{3/2}(F'=9/2)$ repumping transition (see Fig. 2.3). The beam is separated into four beams and expanded by spherical and cylindrical telescopes to create the transverse and longitudinal 2D-MOT beams. The transverse beams have an elliptical cross section ($1/e^2$-diameters: $27.5\,\text{mm}$ and $55\,\text{mm}$), are circularly polarized and retro-reflected by right-angled prisms, which preserve the helicity of the beams. The power losses in the surface of the glass cell and the prisms weaken the power of the retro-reflected beams by $\sim 17\%$ (the loss contribution of the absorption by the vapor is negligible due to the high laser power). This power imbalance is compensated by shifting the position of the magnetic field zero. The longitudinal beams are linearly polarized and have a circular cross section ($1/e^2$-diameter: $27.5\,\text{mm}$). 75% of the fiber output power is used for the transverse beams, 25% for the longitudinal beams. The intensity ratio between pushing and retarding beam along the atomic beam axis is ~ 6 (for reasons explained below).

The optimized values of the essential parameters of the 2D-MOT are displayed in Tab. 2.2. With these parameters a ^{40}K-MOT capture rate of $\sim 1.4 \times 10^9$ atoms/s is obtained.

2.6 ^{6}Li-^{40}K dual-species MOT

Previously, several groups have studied samples of two atomic species in a magneto-optical trap [145, 146, 147, 148, 89, 43, 44]. We describe here the implementation of our ^{6}Li-^{40}K dual-species MOT. Its characterization and a study of collisions between atoms of the different species is presented in chapter 3.

2.6. ^{6}Li-^{40}K dual-species MOT

	^{40}K 2D-MOT
P_{fiber} [mW]	450
$\Delta\omega_{\text{cool}}$ [Γ]	-3.5
$\Delta\omega_{\text{rep}}$ [Γ]	-2.5
$I_{\text{rep}}/I_{\text{cool}}$	1/2
$I_{\text{push}}/I_{\text{ret}}$	6
$\partial_x B, \partial_y B$ [G/cm]	11
K vapor pressure [mbar]	2.3×10^{-7}

Table 2.2: Optimized values for the parameters of the ^{40}K 2D-MOT, yielding a ^{40}K-MOT capture rate of $\sim 1.4 \times 10^9$ atoms/s. The definition of the symbols is given in the text. The natural linewidth of ^{40}K is $\Gamma/(2\pi) = 6.04$ MHz.

2.6.1 Principle of a MOT

In a magneto-optical trap six counter-propagating red-detuned overlapping laser beams cool and magneto-optically confine atoms in a magnetic quadrupole field around its zero [135]. MOTs for alkali atoms require laser light of two frequencies, namely the cooling and the repumping frequency. The latter ensures that the atoms stay in the (almost-) cycling transition used for cooling. Typically the repumping light has a much lower power than the cooling light as the atoms principally occupy the states belonging to the cooling transition. For ^{6}Li, however, the power of the repumping light needs to be relatively high, since ^{6}Li has a very small hyperfine structure in the excited-state manifold (of the order of the linewidth). When laser cooled, ^{6}Li atoms thus very likely quit the cooling transition. Therefore, the repumping light needs to contribute to the cooling process. As a consequence it needs to be present in all six directions with the same polarization as the cooling light. Therefore, we use bichromatic MOT-beams containing both cooling and repumping frequencies. We adapt the same strategy also for ^{40}K.

2.6.2 Experimental setup

Light for the dual-species MOT is derived from two bichromatic laser beams, containing each a cooling and a repumping frequency, which are provided by two separate optical fibers originating from the respective laser systems. The beams are superimposed using a dichroic mirror and then expanded by a telescope to a $1/e^2$-diameter of 22 mm. All subsequent beam reflections are realized by two-inch sized broadband mirrors (Thorlabs, ref. BB2-E02-10). The beam is separated by three two-inch sized broadband polarization cubes (Lambda Optics, ref. BPB-50.8SF2-550) into four arms that form a partially retro-reflected MOT, in which only the vertical beam pair is composed of independent counter-propagating beams. Each retro-reflected MOT beam is focused with a lens in front of the retro-reflecting mirror, in order to increase the intensity and therefore compensate for the losses in the optics and the light absorption by the trapped atoms. The distance between the retro-reflecting mirror and the lens has to be chosen such that the retro-reflected beam is focused at a distance

$$x = \frac{f - \sqrt{P_1/P_2}\, m}{\sqrt{P_1/P_2} - 1} \tag{2.2}$$

from the lens, where f is the focal length of the lens, m its distance from the MOT and P_1, P_2 the power of the incoming and retro-reflected MOT-beam at the position of the MOT. In our setup, we have $P_1/P_2 \sim 1.15$ and we use a lens of focal length 10 cm. The desired focusing is achieved for a distance of ~ 11 cm between the lens and the retro-reflecting mirror.

The distribution of the light power over the MOT beams is independently adjusted for the two wavelengths using a pair of custom-made wave plates, placed in front of each broad-band splitting cube as illustrated in Fig. 2.10. The wave plate pair consists of a $\lambda/2$ plate of order 4 for the wavelength 767 nm and a $\lambda/2$ plate of order 4 for the wavelength 671 nm. To a very good approximation each of these wave plates can turn the polarization direction for one wavelength without affecting the polarization for the other one (since it is $4.5 \times 767 \approx 5 \times 671$ and $4.5 \times 671 \approx 4 \times 767$). The circular polarization of the MOT beams is produced by first order $\lambda/4$ plates for 767 nm, which work sufficiently well also for 671 nm. All four frequency components thus have the same circular polarizations in each of the six MOT beams. A mechanical shutter is placed in the focus of the telescope allowing the production of a total extinction of the MOT light in addition to the partial and fast switching by the AOMs.

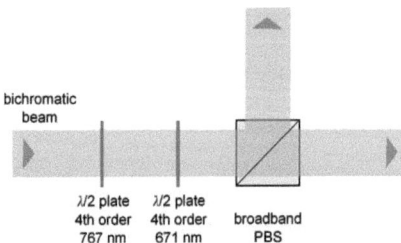

Figure 2.10: Optical setup for the separation of a bichromatic (767 nm+671 nm) linearly polarized beam into two beams. The custom-made wave plates allow the adjustment of the light power in the two beams independently for the two wavelengths. This scheme is employed to create the six MOT beams with the desired powers.

2.7. Magnetic trapping   35

The bichromatic beam for the ^{40}K-MOT has a total power of $P_{\text{fiber}} = 220\,\text{mW}$ and its frequencies are red-detuned by $\sim 3\,\Gamma$ from the $4S_{1/2}(F = 9/2) \to 4P_{3/2}(F' = 11/2)$ cooling and by $\sim 5\,\Gamma$ from the $4S_{1/2}(F = 7/2) \to 4P_{3/2}(F' = 9/2)$ repumping transition (see Fig. 2.3). The intensity of the cooling light is ~ 20 times bigger than that of the repumping light. The bichromatic beam for the ^{6}Li-MOT has a total power of $P_{\text{fiber}} = 110\,\text{mW}$ and its frequencies are red-detuned by $\sim 5\,\Gamma$ from the $2S_{1/2}(F = 3/2) \to 2P_{3/2}(F' = 5/2)$ cooling and by $\sim 3\,\Gamma$ from the $2S_{1/2}(F = 1/2) \to 2P_{3/2}(F' = 3/2)$ repumping transition (Fig. 2.3). The power of the cooling light is ~ 5 times bigger than that of the repumping light.

The magnetic field for the dual-species MOT is created by a pair of coils in anti-Helmholtz configuration. Each coil consist of 6×14 turns of 4 mm thick copper wire of circular cross section which has a hole of 2.5 mm diameter in its center to allow for efficient water cooling. The inner and outer coil diameters are 6.5 cm and 18 cm, respectively. The two coils are separated by 13.4 cm and their total electric resistance (when connected in series) is $0.178\,\Omega$. They create an axial magnetic field gradient of $0.936\,\text{G}/(\text{cm\,A})$. For the optimum dual-species MOT operation the axial magnetic field gradient is chosen to be $\partial_z B = 8\,\text{G/cm}$. This gradient yields an optimum atom number for the ^{40}K-MOT.

The optimum parameters, which lead to atom numbers of $N_{\text{single}} \sim 8.9 \times 10^9$ in the ^{40}K-MOT and $N_{\text{single}} \sim 5.4 \times 10^9$ in the ^{6}Li-MOT, are displayed in Tab. 2.3 together with the characteristics of the MOTs (in dual-species operation, the atom numbers only slightly change due to the additional interspecies collisions to $N_{\text{dual}} \sim 8.0 \times 10^9$ in the ^{40}K-MOT and $N_{\text{dual}} \sim 5.2 \times 10^9$ in the ^{6}Li-MOT). The $(1 - 1/e)$-loading times of the MOTs are $\sim 5\,\text{s}$ for ^{40}K and $\sim 6\,\text{s}$ for ^{6}Li.

2.7 Magnetic trapping

The phase space density which can be achieved in a MOT is limited by the permanent absorption and reemission of photons. In order to reach the quantum degenerate regime, a further increase in phase space density is required, which can only be accomplished in a non-dissipative trap. To achieve this, we transfer the atoms from the MOT to a magnetic trap, which is conservative. Once loaded into this trap the atoms are first transported to another chamber by moving the magnetic trapping potential using time-varying currents through a fixed coil-assembly. Then, the atoms are evaporatively cooled inside the magnetic trap in order to achieve the phase space density necessary for quantum degeneracy.

Another very common conservative trap in which quantum degeneracy can be

	^{40}K-MOT	^{6}Li-MOT
P_{fiber} [mW]	220	110
$\Delta\omega_{\text{cool}}$ [Γ]	-3	-5
$\Delta\omega_{\text{rep}}$ [Γ]	-5	-3
$\Gamma/(2\pi)$ [MHz]	6.04	5.87
I_{cool} per beam [I_{sat}]	13	4
I_{sat} [mW/cm^2]	1.75	2.54
$I_{\text{rep}}/I_{\text{cool}}$	1/20	1/5
$\partial_z B$ [G/cm]	8	8
N_{single} [$\times 10^9$]	8.9	5.4
N_{dual} [$\times 10^9$]	8.0	5.2
n_c [$\times 10^{10}$ at./cm^3]	3	2
T [μK]	290	1400
D_c [$\times 10^{-7}$]	1.2	1.3

Table 2.3: Characteristic parameters of the dual-species ^{6}Li-^{40}K-MOT. n_c is the atomic density in the MOT center, T the temperature of the atoms and $D_c = n_c \Lambda^3$ the peak phase space density with the thermal de Broglie wavelength $\Lambda = \sqrt{2\pi\hbar^2/(mk_B T)}$. The definition of the other symbols can be found in the text.

achieved is the optical dipole trap. However, due to its limited trapping volume, which results in small transfer efficiencies for the atoms from the MOT, we decided to use a magnetic trap for the realization of the evaporative cooling procedure.

Since magnetic trapping is one of the key components of our experimental setup we first describe its principle. Then we describe the experimental sequence for the atom transfer from the MOT to the magnetic trap, the magnetic transport and the planned evaporative cooling inside an optically plugged magnetic quadrupole trap. The implementation of the latter is still in progress, so we will restrict the description of this trap to the constructed parts.

2.7.1 Principle of magnetic trapping

Due to their unpaired electron neutral alkali atoms have a sufficiently large magnetic dipole moment (of the order of a Bohr magneton) to be magnetically trapped. This was first demonstrated in 1985 with sodium atoms [149]. The interaction energy of the magnetic dipole of an atom with an external magnetic field is given by

$$E(\boldsymbol{r}) = -\boldsymbol{\mu}_{\text{m}} \cdot \boldsymbol{B}(\boldsymbol{r}), \qquad (2.3)$$

where $\boldsymbol{\mu}_{\text{m}}$ is the magnetic moment of the atom. It results in a force that drives the atom toward a minimum or a maximum of the magnetic field depending on the orientation

2.7. Magnetic trapping

of $\boldsymbol{\mu}_m$ with respect to the field direction. In order to trap the atom, the magnetic field has to have an extremum of field amplitude. Due to Maxwell's equations [150] no static magnetic field maximum exists in free space. Minima of the field amplitude, however, do exist and thus the atom can be trapped if it is in a state with a magnetic moment antiparallel to the field direction (low field seeker). The condition for its trapping, however, is that its magnetic moment can follow the direction of the magnetic field when it changes sign. Thus, to remain trapped, the rate of change of the field direction experienced by the atom must be much smaller than the Larmor frequency ω_L, i.e. $|\mathrm{d}(\boldsymbol{B}/B)/\mathrm{d}t| = |(\boldsymbol{v} \cdot \nabla)\boldsymbol{B}/B| \ll \omega_L = \mu_m B/\hbar$. If this adiabatic condition is violated, the atom changes its internal state eventually to a high field seeking state, and will be repelled from the trap. Such "Majorana spin-flip" transitions mostly occur at the center of the trap, where the magnetic field changes its direction.

For small magnetic fields, the linear Zeeman regime holds, and the energy levels of the atom in the magnetic field are $E(m_F) = g_F m_F \mu_B B$, where g_F is the Landé g-factor of the hyperfine state, m_F the quantum number of the z-component of the angular momentum $\boldsymbol{F}$ of the atom and $\mu_B \approx h \times 1.4\,\mathrm{MHz/G}$ the Bohr magneton. The states with $g_F m_F > 0$ are the trappable low field seeking states. The dependence of the energy levels on the magnetic field for arbitrary magnetic fields can be calculated using the Breit-Rabi formula [151]. They are depicted for the hyperfine ground states of ^{6}Li and ^{40}K in Fig. 2.11 (the low-field seeking states have a positive slope in the figure). The extension of the linear Zeeman regime depends on the strength of the hyperfine interaction of the atom. For ^{6}Li it extends to $\sim 27\,\mathrm{G}$ and for ^{40}K it extends to $\sim 357\,\mathrm{G}$. For the stretched states $m_F = \pm\max(F)$ the linear Zeeman regime extends to large magnetic fields.

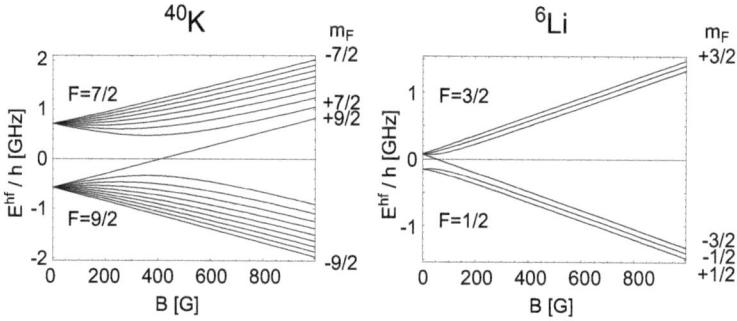

Figure 2.11: Energy levels of ^{40}K and ^{6}Li in a magnetic field. Note that the hyperfine structure of ^{40}K is inverted, with $F = 9/2$ being lower in energy than $F = 7/2$. This particularity of ^{40}K allows stable magnetic trapping of two different spin states.

In our setup we use a magnetic trap in both the MOT chamber and the science cell. Both magnetic traps are created by a coil pair in anti-Helmholtz configuration, which approximately generates a magnetic quadrupole field of the form

$$\boldsymbol{B}(\boldsymbol{r}) = \frac{1}{2}\begin{pmatrix} -B'x \\ -B'y \\ 2B'z \end{pmatrix}, \qquad (2.4)$$

where B' is the gradient of the magnetic field in the axial direction z. This axially symmetric quadrupole field yields an adiabatic potential

$$V(\boldsymbol{r}) = \frac{1}{2} g_\mathrm{F} m_\mathrm{F} \mu_\mathrm{B} B' \sqrt{x^2 + y^2 + 4z^2}. \qquad (2.5)$$

It is linear in r along the coordinate axes and traps the atoms around its minimum at $r = 0$. Since there the magnetic field changes direction, atoms passing too close to the trap center will undergo a spin-flip. The trap center thus represents a "hole", whose effective size depends on the velocity of the atoms. For an atom with velocity v the adiabaticity condition yields an effective hole size (along the z-direction) of $z_0 = \sqrt{\hbar v/(\mu_\mathrm{m} B')}$ [9]. The loss rate induced by the hole is given by the flux of atoms through its surface (an ellipsoid of radii z_0 and $z_0/\sqrt{2}$), which is approximately given by $\Gamma = \hbar/(ml^2)$ [152], where m is the atomic mass and l the trap extension. This rate is small as long as $z_0 \ll l$. For a ^{6}Li-cloud with temperature $T = 1\,\mathrm{mK}$ and a magnetic field gradient of $B' = 100\,\mathrm{G/cm}$ it is $z_0 \sim 10\mu\mathrm{m}$ and $l \sim 2.7\,\mathrm{mm}$, resulting in $\Gamma \sim 1.4\times 10^{-3}\,\mathrm{s}^{-1}$. For a ^{40}K-cloud with temperature $T = 300\,\mu\mathrm{K}$ and $B' = 100\,\mathrm{G/cm}$ it is $z_0 \sim 5\mu\mathrm{m}$ and $l \sim 1\,\mathrm{mm}$, resulting in $\Gamma \sim 1.9\times 10^{-3}\,\mathrm{s}^{-1}$. The Majorana losses are thus negligible when the atoms have a temperature which corresponds to the temperature in the MOT.

When an atom cloud is evaporatively cooled inside a quadrupole magnetic trap, however, the relative size of the hole with respect to the cloud size increases and will eventually become so significant that the evaporation stops. The hole thus needs to be plugged. This can be either done with an additional bias magnetic field (as in the Ioffe-Pritchard trap [153] or the time-averaged orbiting potential (TOP) trap [152]) or by using the repulsive force of a blue detuned laser beam [154], which is pointed at the trap center. This "optically plugged magnetic trap" achieves tight confinement allowing for efficient evaporative cooling and it yields a large optical access. We started the implementation of this trap, whose current setup is described in Sec. 2.7.5.

2.7.2 Transfer from the MOT to the magnetic quadrupole trap

As soon as the dual species MOT is loaded with a sufficiently large number of atoms, we transfer the atom cloud to the magnetic quadrupole trap, which is created by the coil pair which is also used for the MOT. For an efficient transfer the atom cloud first needs to be compressed and it needs to be polarized to magnetically trappable states. This requires two stages termed the "compressed MOT"- and the "optical pumping"-stage, which are executed in immediate succession before the magnetic field of the trap is switched on.

Compressed MOT

The atomic clouds are compressed by temporarily changing the parameters of the MOTs during the transfer. We optimized the compressed MOT stage for ^{40}K, since we are essentially interested in a large atom number of ^{40}K in the science cell for evaporative cooling. It was found optimum to ramp down the magnetic field gradient to zero within 3.5 ms, to increase the cooling frequency detuning from $-3\,\Gamma$ to $-5.5\,\Gamma$ and to decrease the repumping frequency detuning from $-5\,\Gamma$ to $-\Gamma$. The decrease of the repumping frequency detuning leads to a more efficient pumping of the atoms to the hyperfine ground state $F=9/2$, and the increase of the cooling frequency detuning leads to a decrease of the light pressure of the rescattered photons in the cloud. The light intensities are kept unchanged. After this sequence, the temperature decreases from $\sim 290\,\mu$K to $\sim 200\,\mu$K and the peak density increases from $\sim 3 \times 10^{10}$ at./cm^3 to $\sim 5 \times 10^{10}$ at./cm^3. No significant loss in atom number is observed. For the magnetic field gradient imposed by the value chosen for ^{40}K the optimum compressed MOT stage for ^{6}Li lasts 5.5 ms. During this time, the cooling light intensity is ramped down to zero and its frequency detuning is decreased from $-5\,\Gamma$ to $-3\,\Gamma$. The repumping light intensity is set to 30% of its initial value and its frequency is unchanged. The temperature of the ^{6}Li cloud descreases from $\sim 1.4\,$mK to $\sim 1\,$mK and the peak density increases from $\sim 2 \times 10^{10}$ at./cm^3 to $\sim 3.5 \times 10^{10}$ at./cm^3.

Higher densities of up to $\sim 10^{11}$ at./cm^3 could be obtained for both clouds by increasing the magnetic field gradient during the compressed MOT stage, but this led to a significant atom loss of about a factor 4 in both clouds. Since the phase space density of the clouds in the final cell after the transport was found to be only slightly affected by the parameters of the compressed MOT stage, we optimized its parameters for the maximum number of atoms in the final cell, which was obtained for the parameters given above.

Optical pumping

In order to avoid loss of atoms which occupy high field seeking states or states which are susceptible to undergo spin relaxation, all atoms are optically pumped into the stretched states before the magnetic trap is switched on. This is accomplished by shining a pair of counterpropagating near-resonant circular polarized light beams onto the atoms. For ^{40}K the light beams are bichromatic containing a principal and a repumping frequency, both on resonance. The beam pair has a peak intensity of $\sim$ 90 I_{sat} per beam (20% of the beam intensity consisting of repumping light) and the light pulse has a duration of 5 μs. For ^{6}Li the beams consist of only one single frequency, which is detunned by $-3\,\Gamma$ from the repumping transition. The peak intensity is $\sim$ 30 I_{sat} per beam and the light pulse has a duration of 200 μs. Due to the small ground-state hyperfine structure of ^{6}Li the beams are detuned by only $+36\,\Gamma$ from the principal transition, leading to a significant number of absorptions for atoms occupying the principal state. The high intensities used in our optical pumping schemes ensure that the optical pumping beams are not significantly attenuated by absorption from the atoms, resulting in a homogeneous exposure of all atoms in the cloud.

The optical pumping beam pairs for ^{6}Li and ^{40}K are provided by two optical fibers originating from the respective laser systems. They are combined using a dichroic mirror and subsequently expanded by a telescope to a $1/e^2$-diameter of 1.1 cm. The combined beam is then split into two parts using a broadband polarization cube, and both parts are shone on the MOTs from opposite directions (in order to avoid an acceleration of the atoms along the beam direction). The circular polarization of the beams is created using wave plates for ^{40}K. In order to define the quantization axis for the atomic spins a small bias magnetic field of 2 G along the beam direction is applied.

The optical pumping stage was found particularly efficient for ^{40}K, as it yields an atom number gain in the magnetic trap of a factor of $\sim$ 10. We achieve transfer efficiencies of nearly 100% for ^{40}K and $\sim$ 30% for ^{6}Li. For ^{6}Li the efficiency is smaller, since ^{6}Li has only one single simultaneously trappable spin state (as opposed to five for ^{40}K), such that depolarizing events, which are caused by emission and reabsorption of randomly polarized photons, more likely lead to trap loss. For ^{40}K the optical pumping stage was found to lead to a significant heating of the cloud which can be attributed to the high light intensity used, since it could be entirely cancelled by decreasing the intensity. For an optical pumping stage which does not induce heating the maximum transfer efficiency is $\sim$ 70%.

After the optical pumping stage, it is very important that stray light originating from the laser sources is completely extinct, as it might provoke spin-flip transitions to untrapped states induced by optical excitations followed by spontaneous emissions. In

2.7. Magnetic trapping

our experiment we therefore switch off all near-resonant light beams using AOMs and mechanical shutters during periods in which atoms are magnetically trapped. We also detune the light frequencies far from resonance. Due to the AOM's fast switching times (of the order of hundreds of ns), they allow one to quickly attenuate the light beams. Since the attenuation is not perfect (only of the order of 10^{-4}), mechanical shutters are employed to accomplish a complete extinction. This is most efficiently done by placing the shutters in front of the optical fibers which transfer the light beams to the experimental table. The shutters have typical closing times of the order of 100 μs, when placed in the focus of a beam.

2.7.3 Magnetic transport

After both atomic species have been loaded into the magnetic trap, they are transported from the MOT chamber into the science cell by means of magnetic forces. This transport allows us to perform the evaporative cooling stage in the more favorable vacuum environment of the science cell. Furthermore the evaporative cooling can be performed more efficiently there due to the better mechanical access to the atoms which allows placing the magnetic quadrupole coils very close to the atoms, thus creating a steep confinement.

The magnetic transport mechanism is based on shifting the quadrupole potential of the magnetic trap adiabatically such that the atomic cloud follows. Two different possibilities exist to accomplish this: either, the quadrupole coil pair of the magnetic trap is moved mechanically [155], or a series of overlapping fixed coils is used, which create the moving trapping potential by applying time-varying currents [131]. We decided to implement the latter configuration, since it yields a high reproducibility due to minimized mechanical vibrations and less stringent geometric constraints.

The basic mechanism of the fixed coil configuration is the following. In principle, an atom cloud can be moved from the center of one quadrupole pair to the center of its neighboring (partially overlapping) quadrupole pair by simply increasing the current in the neighboring quadrupole pair and decreasing it in the first one. During this transport the atom cloud remains trapped inside a quadrupole potential, since two partially overlapping quadrupole coil pairs with the same current orientations create a quadrupole potential again. During this transfer, however, the aspect ratio of the quadrupole potential changes from $A = \frac{\partial B_x}{\partial x} / \frac{\partial B_y}{\partial y} = 1$ for the initial position to $A > 1$ (elongated cloud) back to $A = 1$ for the final position. Repeating this procedure for the remaining coil pairs would thus appreciably modulate the potential geometry which leads to severe heating of the atom cloud. This can be avoided by continuously applying current to three coil pairs in the transport process, which avoids that the aspect ratio changes back to $A = 1$ after each coil pair, but rather stays at a constant value. A change of aspect ratio is thus only created at the beginning and at the end of the entire transport. Furthermore, the implication of three coils allows keeping the field gradients constant during the major part of the transport.

The design of our magnetic transport is based on the one used for its first demonstration [131]. In the following we present a brief description of the setup, for details we refer the reader to the thesis of Thomas Salez [156]. A sketch of the coil assembly is shown in Fig. 2.12. The atom cloud is transported over a total distance of 50 cm employing 14 partially overlapping quadrupole coil pairs. The transport path contains a 90° corner ("elbow") at a distance of 30 cm in order to give additional optical access along the final transport direction and to reduce trap loss due to collisions with background atoms from the MOT chamber. The corner prohibits these atoms to reach the atoms in the science cell in ballistic flight. Except for the coils at the beginning and the end of the transport, all coils of the assembly are identical. Since the overlap of the MOT coils and the first transport coil is limited by the MOT beams an aspect ratio much larger than in the rest of the transport would be created if the cloud was moved toward the first transport coil according to the procedure described above. This excessive aspect ratio can be avoided by the use of an additional so-called "push coil", which is placed opposite to the transport coil assembly with its axis oriented parallel to the transport direction (see Fig. 2.12). It permits to shift the trap center toward the coil assembly.

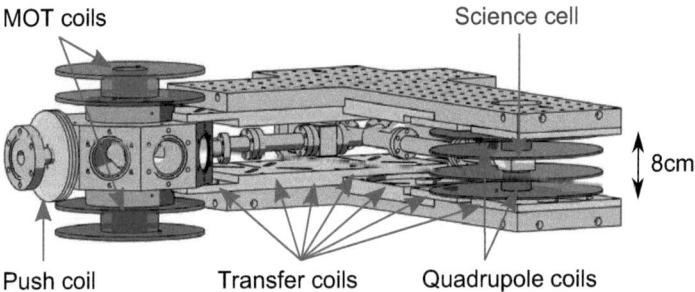

Figure 2.12: Schematics of the magnetic transport coil assembly. Time-varying currents through 14 partially overlapping quadrupole coil pairs (MOT coils, transfer coils and quadrupole coils) and a single "push" coil transport the atom cloud from the MOT chamber to the science cell. The transport path has a total length of 50 cm and contains a 90° corner to allow additional optical access to the science cell and to increase the vacuum quality there.

2.7. Magnetic trapping

The push coil is placed at a distance of 11.1 cm from the MOT, it was wound by ourselves. It consists of 8 × 4 turns of copper cable with a circular cross section of 2 mm diameter. The inner and outer diameters of the coil are 9.1 cm and 13.2 cm, respectively. It creates a magnetic field of 0.25 G/A at the position of the MOT. The coil resistance was measured to be 0.06 Ω. The transport coils have been manufactured by the company Oswald. They have the shape of a flat disc, so as to permit overlapping of two coils without appreciably changing their distance to the transport path. The coils are arranged in two layers above and below the transport path (see Fig. 2.12). Each coil consists of 47 × 2 turns with inner and outer diameters of 2.3 cm and 13.3 cm, respectively. The inner coils are separated by 7-8 cm. They create an axial magnetic field gradient of 3.2 G/(cm A). Their resistance was measured to be 0.36 Ω. All transport coils are mounted on separate aluminum holders, made in our workshop. Water from a cooling system flowing through the holders ensures operation at constant temperature. Eddy currents are suppressed by numerous slits in the holders. A photograph of the transport coil assembly is shown in Fig. 2.13.

Figure 2.13: Photograph of the transport coil assembly before its integration into the experimental system.

A computer program was written [156] which calculates the required values of the current in each coil as a function of the atom cloud position. The temporal dependence of the currents is then determined by the desired dynamics of the cloud position, which can be chosen independently. Four different unipolar current sources are used for the transport, which are controlled by the analog voltages provided from the experimental control. The current sources are controlled in constant voltage mode rather than constant current mode as the former allows quicker switching times, minimizing the time delay between the demanded and the obtained currents. One current source is used for the pushing coil and the others are used to drive three adjacent coil pairs in parallel. For an efficient use of the current sources, a control logic was installed which switches their connections to the appropriate set of three coils during the transport. The switching is realized with a series of metal-oxide semiconductor field-effect transistors (MOSFET) (Farnell, ref. IXFN200N10P), which open or close the required connections on demand.

The switching of coil currents creates an induction voltage $U_\mathrm{ind} = -L\dot{I}$, where L denotes the self inductance of the coil, which falls off at the junction of the switch. In order to avoid that this voltage destroys the MOSFETs, we installed varistors between their source and drain connectors, which dissipate the energy of the magnetic field during the switching. The conducting voltage of the varistors needed to be chosen lower than the break down voltage of the MOSFETS. Also the control logic needs to be protected from the induction voltage, which is realized with the help of optocouplers.

The transport path passes through a series of CF16- and homemade vacuum tubes of 1 cm diameter (see Fig. 2.12). Those tubes had been designed to allow most of the magnetically transported atoms to pass through and to sufficiently limit their conductance to have a low enough pressure in the science cell. However, we realized, that the transport efficiency significantly suffers from the small tube diameters. Those make the alignment of the transport coil assembly with respect to the tubes very critical. It is thus helpful to run high currents through the coils in order to achieve a substantial confinement of the atom cloud. This also allows the reduction of the transport time and thus the losses due to collisions with background atoms. With the initial design, we could achieve transport efficiencies of only 5%. We had to buy additional power supplies and place an additional coil pair between the MOT and the first transfer coil pair in order to create a stronger confinement, which finally resulted in transfer efficiencies of $\sim 15\%$.

2.7.4 Magnetic quadrupole trap of the final cell

The magnetic transport sequence ends when the atoms arrive in the magnetic quadrupole trap of the final cell. In this trap the atoms will be evaporatively cooled until they are sufficiently cold and compressed to be transferred into an optical dipole trap. The atom loss due to Majorana spin flips during the evaporative cooling will be avoided by the presence of an optical plug which repels the atoms from the region

2.7. Magnetic trapping

where the spin flips occur. We present here the specifications of the final quadrupole trap, the installation of the optical plug is described in the next section.

The coil pair for the magnetic trap in the science cell was wound by ourselves and it consists of 4×19 turns of 4 mm thick copper wire of circular cross section which has a circular hole of 2.5 mm diameter in its center to allow for efficient water cooling. The inner and outer coil diameters are 3.4 cm and 19.5 cm, respectively. The two coils are separated by 3.65 cm, leaving a distance of 3.3 mm between the coils and the walls of the science cell. They create an axial magnetic field gradient of 3.75 G/(cm A). We will soon replace this coil pair by a new one, which is manufactured by the company Oswald. The mount for this new coil pair will also support a coil pair which can create a strong bias field with a high precision which will allow tuning atomic interactions by means of Feshbach resonances in an optical trap.

The maximum trap depth of a quadrupole trap is defined by the maximum achievable radial gradient and the geometric boundary of the trap, which is defined by the inner walls of the vacuum chamber when the coils are placed outside the chamber. The coil pair of the science cell can create a maximum radial gradient of 280 G/cm (for a coil current of 150 A) and the distance between the trap center and the chamber walls is ~ 1 cm, resulting in a maximum achievable trap depth of ~ 18 mK (for comparison, the coil pair of the MOT chamber can create a maximum radial gradient of 70 G/cm for the same coil current and the distance between the trap center and the chamber walls is ~ 2.2 cm, resulting in a maximum achievable trap depth of ~ 10 mK).

In order to allow fast switching times of the magnetic field of the quadrupole trap, we installed an insulated gate bipolar transistor (IGBT) (Mitsubishi Electric, ref. CM600HA-24H), which can entirely interrupt a current flow of 150 A through the coil pair within ~ 1 ms. The switching speed of IGBTs is superior to that of MOSFETs: for example, the MOSFET-model which we employ for the switching of the transport coils, would require ~ 13 ms for switching off the above current flow in the magnetic trap coils. In order to protect the IGBT from induction currents produced during the switching, we connect a varistor in parallel to its junction, which dissipates the energy of the magnetic field. In addition, the computer control of the IGBT is electronically isolated using opto-couplers.

2.7.5 Optical plug

The "optically plugged magnetic trap" configuration was first implemented in 1995 and allowed for the first realization of a Sodium Bose-Einstein condensate [154]. It generates a steep, linear confinement, which leads to high elastic collision rates and thus to efficient evaporative cooling. Due to its complicated geometry, it is, however, not very practical for quantitative studies of atom clouds and we intend to employ it only in order to cool the atoms. Once sufficiently cold, the atom cloud will be transferred into an optical dipole trap. It has been demonstrated by the Amsterdam group [44], that the mixture ^{6}Li-^{40}K can be efficiently cooled inside this trap.

The installation of the optical plug is still in progress. We give here only a brief description of the constructed parts. Then, we give an outlook on the planned evaporative cooling strategy.

We have purchased an intracavity-doubled Nd:YVO4 laser (Coherent, ref. Verdi-V12) of 532 nm wavelength and 12 W output power from which the light for the optical plug is derived. Heating of the atoms due to photon scattering will be suppressed due to the large detuning of the laser beam with respect to the resonance frequency of both atomic species, and due to the repulsion of the atoms from regions with high intensity. We have installed an optical system which focuses 7 W of power to a size of $\sim 20\,\mu$m. The repulsive barrier induced by the laser beam has a height of ~ 1 mK for the two species. For an axial magnetic field gradient of 400 G/cm, the magnetic field at the trap bottom is ~ 0.4 G. The optical plug can be switched off by an air-cooled 110 MHz-AOM (AA optoelectronic, ref. MCQ110-A2 VIS), which is driven by a homemade voltage-controlled oscillator, whose output is amplified by an rf-amplifier (Minicircuits, ref. ZHL-5W-1). The beam path from the laser to the science cell is protected from dust particles by a Thorlabs cage system, which in addition gives a high mechanical stability to the system, reducing the pointing noise of the beam focus. The second last mirror in front of the science cell is controlled by an electronic circuit (Thorlabs, ref. TST001) using two motor actuators (Thorlabs, ref. ZST13), and allows for the adjustment of the plug position. The plug beam is aligned to the symmetry axis of the magnetic quadrupole trap. Its focus will be continuously imaged to the same CCD-camera as the trapped atoms in order to verify its correct alignment at any time. Since we have not achieved evaporative cooling, we could not verify the functioning of the optical plug, yet.

2.7.6 Evaporative cooling

As discussed in the introduction of this book, evaporative cooling of fermionic species is more difficult than for bosonic species, due to the suppression of s-wave collisions for identical particles. While odd partial wave collisions, such as p-wave collisions, are allowed, they are suppressed at low temperatures (typically below $T \approx 6$ mK for ^{6}Li and $T \approx 300\,\mu$K for ^{40}K [157]), because of the angular momentum barriers. Therefore,

2.7. Magnetic trapping

high rates for elastic collisions are only possible at low temperatures if the colliding atoms are either of two different species or in at least two different internal states.

Three possibilities exist to evaporatively cool the mixture ^{6}Li-^{40}K. Either, (1) the mixture is evaporatively cooled relying only on interspecies collisions, or (2) each atomic species is prepared in two different internal states and evaporatively cooled separately, or (3) one species is prepared in two different spin states and evaporatively cooled, sympathetically cooling the other species along.

The first possibility would be relatively inefficient due to the different mass of both species and the small cross section for ^{6}Li-^{40}K collisions. For this kind of evaporation to be maximally efficient, an equal number of atoms in each species is required and the atoms should be prepared in their fully-stretched states, which suppresses spin-relaxation. The different mass of the colliding particles leads to longer rethermalization times of about a factor of 2 [44] as compared to colliding particles of the same mass. The collisions, which take place in the triplet channel, have a s-wave scattering length of $a = 64.41\,a_0$ [91], leading to a collision cross section of $\sigma_{\text{LiK}} = 1.5 \times 10^{-10}\,\text{m}^2$. Furthermore, the different initial temperatures of the ^{6}Li-MOT and the ^{40}K-MOT will lead to a significant heating of ^{40}K.

The second possibility cannot be realized in a magnetic trap, since ^{6}Li cannot be magnetically trapped in two different spin states which are stable against inelastic collisions. This is, because the hyperfine structure of ^{6}Li (see Fig. 2.11) implies that at least one of the two different magnetically trappable spin states belongs to the higher F state manifold. Since collisions between any two different trappable spin states of ^{6}Li can lead to two states in the lower F state manifold spin relaxation can occur [136]. In order to take advantage of the large trapping volume of magnetic traps, evaporative cooling inside this trap is preferable, thus we do not further consider this possibility.

The third possibility seems to be the most efficient. It has been demonstrated that ^{40}K can be trapped magnetically in the two different spin states $|F, m_F\rangle = |9/2, 9/2\rangle$ and $|9/2, 7/2\rangle$ with suppressed spin relaxation [21]. This suppression is a consequence of the inverted hyperfine structure of ^{40}K making the state with the larger F ($F=9/2$) the hyperfine state with the lowest energy. Therefore, evaporatively cooling ^{40}K in a magnetic trap is possible. The s-wave scattering length for the intraspecies collisions between the above-mentioned spin states is $a \sim 170\,a_0$ [158], leading to a collision cross section of $\sigma_{\text{KK}} = 1 \times 10^{-9}\,\text{m}^2$, which is nearly an order of magnitude larger than that for intraspecies ^{6}Li-^{40}K collisions. The evaporation of ^{40}K using this method is thus more efficient when the evaporation is based on intra- rather than interspecies collisions. The sympathetic cooling of ^{6}Li will also be efficient if a small cloud of ^{6}Li is loaded into the trap. This has been demonstrated by the Amsterdam group [44].

We plan to apply this cooling technique as well. Special care, however, will need to be taken of the spin-exchange collisions which depolarize the ^{6}Li atoms. Those increase with the density of the ^{40}K cloud, which thus should be kept low. When the cloud is transferred to the optical dipole trap, state purification will be required to allow for the final compression of the sample.

2.8 Diagnostic tools

Several methods exist to probe a cloud of cold atoms. The method we chose in order to probe the atoms in the two different chambers is the absorption imaging technique, which we describe in detail in the subsequent section. For the atoms which are trapped by the MOTs only, we also use other ways for probing which we briefly describe at the end of this section.

2.8.1 Principle of absorption imaging

In absorption imaging, a near-resonant laser beam of very low intensity ($I \ll I_{\text{sat}}$) is shone on the atomic cloud for a short time and its transmission profile is recorded by a charge-coupled device (CCD) camera. Since the atoms absorb the imaging beam, the transmission profile contains information about the atomic distribution. However, useful data cannot directly be extracted from the recorded transmission profile $I_{\text{abs}}(x,y)$ alone. Since the intensity profile of the imaging beam is inhomogeneous and since background ambient light is always present, the transmission profile needs to be normalized. Therefore two more pictures need to be taken: first, a reference picture, in which the intensity profile $I_{\text{ref}}(x,y)$ of the unabsorbed imaging beam is recorded. This can be done by either waiting for the atoms to fall off the field of view or by making them invisible for the imaging beam (see discussion below). Second, a background picture, in which the background signal $I_{\text{bg}}(x,y)$ is recorded in absence of the imaging beam. The normalized transmission profile $T(x,y)$ is then obtained from these three pictures by

$$T(x,y) = \frac{I_{\text{abs}}(x,y) - I_{\text{bg}}(x,y)}{I_{\text{ref}}(x,y) - I_{\text{bg}}(x,y)} \tag{2.6}$$

(see Fig. 2.14). The subtraction of $I_{\text{bg}}(x,y)$ also accounts for the unavoidable dark current rate of the CCD pixels, which may vary with temperature.

The recorded normalized transmission profile is directly related to the atomic density distribution $n(x,y,z)$ inside the cloud. Since the imaging beam has a low intensity,

2.8. Diagnostic tools

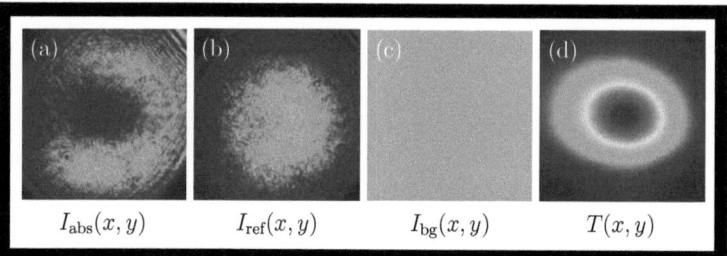

Figure 2.14: Obtaining a normalized transmission profile $T(x,y)$ requires to take three images: (a) an image of the probe beam after passing through the atom cloud, (b) an image of the probe beam in absence of the atoms and (c) a background image. These are processed according to Eq. (2.6) to give the normalized transmission image (d). The picture shows a magnetically trapped cloud of $\sim 3 \times 10^9$ ^{40}K atoms.

its relative transmission through the atom cloud is independent of its intensity and $T(x,y)$ is related to $n(x,y,z)$ by

$$T(x,y) = \exp\left(-\mathrm{OD}(x,y)\right) = \exp\left(-\sigma \int n(x,y,z)\mathrm{d}z\right), \qquad (2.7)$$

where $\mathrm{OD}(x,y)$ is defined as the optical density, and σ is the scattering cross section. The integration is performed along the propagation direction z of the imaging beam. σ is for low beam intensities approximately given by

$$\sigma = C^2 \frac{3\lambda^2}{2\pi} \frac{1}{(1+(2\Delta\omega_{\mathrm{img}}/\Gamma)^2)}, \qquad (2.8)$$

where λ is the wavelength of the atomic transition, $\Delta\omega_{\mathrm{img}}$ the detuning of the imaging laser from resonance and C the Clebsch Gordan coefficient of the transition, which accounts for the departure from the two-level approximation. It is desired to use a closed transition for the imaging, since then σ is given by the two-level atom approximation for which $C = 1$. For ^{6}Li such a closed transition is given by $|F=3/2, m_\mathrm{F}=3/2\rangle \rightarrow |F'=5/2, m_{\mathrm{F'}}=5/2\rangle$ and for ^{40}K by $|F=9/2, m_\mathrm{F}=9/2\rangle \rightarrow |F'=11/2, m_{\mathrm{F'}}=11/2\rangle$. To ensure that one of these transitions is probed, one in principle would have to spin-polarize the atoms before taking the image and one would have to use an imaging beam with σ^+-polarization (in presence of a small bias magnetic field directed along the imaging beam propagation to define a quantization axis for the atomic spins).

When the atoms are not spin-polarized before taking the image, the distribution of the atoms over the possible spin states is unknown. The best estimate for C^2 in Eq. (2.8) is therefore given by the average of the squared Clebsch Gordan coefficients of all possible transitions, which can be excited by the imaging beam (normalized to the one for the cycling transition). For ^{6}Li the average value is calculated to $C^2 = 0.5$ and for ^{40}K it is $C^2 = 0.4$. When we image the atoms which are trapped in the MOT we use these average Clebsch Gordan coefficients. In general, the biggest uncertainty in the determination of the atomic density using absorption images arises from the estimate of the Clebsch Gordan coefficient. We estimate the uncertainty of the measured atomic density to be 50%.

The atoms which are trapped in the magnetic quadrupole trap are spin polarized in their stretched states. When we image these atoms in the science cell, we apply a small bias magnetic field of ~ 2 G during the imaging pulse (after switching off the magnetic quadrupole trap) whose direction is parallel to the axis of the magnetic quadrupole trap. Since the atoms will adiabatically align their spins to this field, we use $C^2 = 1$.

The bias field for the imaging in the science cell is created by two coils in Helmholtz-configuration. Each coil consists of 10 turns with 11 cm diameter. They are placed in a distance of 16 cm from each other. The resulting magnetic field at the position of the atom cloud is 0.4 G/A.

2.8.2 Evaluation of absorption images

In the previous paragraph we have seen that the measurement of the transmission profile gives access to the density distribution of the atom cloud. In general, all properties of the atom cloud can be inferred from measurements of density distributions by comparing them to the results of models of the atomic gas. We briefly discuss here how to infer the basic quantities characterizing the atom cloud, such as its total atom number, its local atomic density and its temperature. The total number N of atoms is determined by integrating the recorded optical density profile across the two-dimensional image of the cloud

$$N = \int \mathrm{d}x \int \mathrm{d}y \int \mathrm{d}z \, n(x,y,z) = \int \mathrm{d}x \int \mathrm{d}y \, \frac{OD(x,y)}{\sigma}, \qquad (2.9)$$

with the scattering cross section σ given by Eq. (2.8).

The determination of the local atomic density $n(x, y, z)$ from the measured optical density profile $OD(x, y)$ is *a priori* not simple, since $OD(x, y)$ contains information about the integrated atomic density distribution only. In order to have access to $n(x, y, z)$, the optical density profile would therefore, in principle, have to be recorded along two different imaging directions. However, when the atom cloud has cylindrical symmetry (with the symmetry axis along the x-direction), $n(x, y, z)$ can be inferred from a single optical density profile. In this case $n(x, y, z)$ is given by the inverse Abel

2.8. Diagnostic tools

transform of $\mathrm{OD}(x,y)$ [159]

$$n(x,y,z) = n(x,r) = -\frac{1}{\sigma\pi}\int_r^\infty \left(\frac{\partial \mathrm{OD}(x,y)}{\partial y}\right)\frac{dy}{\sqrt{y^2-r^2}}, \qquad (2.10)$$

where $r = \sqrt{y^2 + z^2}$ denotes the distance to the center of the atom cloud. The determination of $n(x,y,z)$ via Eq. (2.10) in general requires a numerical integration. In some cases, however, a numerical integration can be avoided. This is the case when $n(x,y,z)$ can be assumed to have a simple form which allows the calculation of the corresponding optical density profile analytically. Fitting the measured optical density profile to the calculated form thus allows one to directly obtain the parameters which specify $n(x,y,z)$. For example, when the atomic density distribution can be assumed to be a cylindrically symmetric Gaussian (with widths σ_x, σ_{yz} and peak atomic density n_c, centered around the origin), the optical density profile takes the form [160]

$$n(x,r) = n_c e^{-\frac{x^2}{2\sigma_x^2} - \frac{r^2}{2\sigma_{yz}^2}} \quad \Longleftrightarrow \quad \mathrm{OD}(x,y) = n_c \sigma \sigma_{yz}\sqrt{2\pi}\, e^{-\frac{x^2}{2\sigma_x^2} - \frac{y^2}{2\sigma_{yz}^2}}, \qquad (2.11)$$

which is also a Gaussian. This atomic distribution assumption is exact for a classical gas in a cylindrically symmetric harmonic trap, but is also well fulfilled for not too dense clouds trapped in a MOT. When the atomic density distribution can be assumed to be constant over a sphere (with radius R), the optical density profile takes the form [160]

$$n(x,r) = \begin{cases} n_c & \text{for } 0 \leq |y| \leq R(x), \\ 0 & \text{otherwise}. \end{cases} \quad \Longleftrightarrow \quad \mathrm{OD}(x,y) = 2n_c\sigma\sqrt{R(x)^2 - y^2}, \qquad (2.12)$$

with $R(x) = \sqrt{R^2 - x^2}$. This assumption typically holds for clouds trapped in a MOT with large atom numbers. In order to determine the central atomic density in the MOT, we used Eqs. (2.11) and (2.12) to fit the measured optical density profile. For both cases, the fits nearly yielded the same central optical density. In order to allow the determination of the atomic density of more complicated atomic distributions we wrote a computer program, which carries out the numerical integration of Eq. (2.10). This calculation is very sensitive to noise in the optical density profile $\mathrm{OD}(x,y)$, because it contains the derivative of $\mathrm{OD}(x,y)$. The optical density profile OD thus was smoothened before its derivative was calculated. For the MOT, using the full numerical method and the fitting method yielded nearly the same central atomic density, which confirms both methods for the given application.

The temperature of an atom cloud can be determined by the time-of-flight (TOF) method, in which the trapping potential is suddenly switched off and the expansion of the cloud is measured as a function of the flight time t_{TOF}. For an atom cloud with an initial atomic density distribution of approximately Gaussian shape and a Maxwell-Boltzmann velocity distribution, the atomic density distribution maintains its Gaussian shape during a collisionless expansion. The waist σ_i of the Gaussian distribution changes in time according to

$$\sigma_i(t_{\text{TOF}}) = \sqrt{\sigma_i^2(t_{\text{TOF}} = 0) + \frac{k_B T}{m} t_{\text{TOF}}^2}, \qquad (2.13)$$

with $i \in \{x, y, z\}$ and T denoting the temperature of the cloud. The assumption of an initial Gaussian distribution is exact for a classical gas in a harmonic trap. If it is non-Gaussian a long TOF is required, such that the initial size of the cloud can be neglected. Since it is not always possible to expand the cloud to much larger than its initial size due to limited field of view, the temperature of the cloud is most precisely determined by fitting the time-evolution of the cloud size to Eq. (2.13).

2.8.3 Optical setup

Each atomic species requires its own imaging beam, which is provided by a separate optical fiber originating from the respective laser system (see Fig. 2.4). The two imaging beams are superimposed using a dichroic mirror and expanded by a telescope to a $1/e^2$-diameter of 27.5 mm. The bichromatic imaging beam is subsequently divided into two parts by a polarizing broadband beam splitter. One part serves to image the atoms inside the MOT chamber, the other one is used to image the atoms in the final cell. In each part both frequency components have a low intensity of $I_{\text{img}} \sim 0.01 I_{\text{sat}}$ in the beam center and the circular polarizations are prepared with a $\lambda/4$-plate for the potassium wavelength, which works sufficiently well also for lithium. Both beam components are near-resonant with respect to the $4S_{1/2}(F = 9/2) \rightarrow 4P_{3/2}(F' = 11/2)$ and the $2S_{1/2}(F = 3/2) \rightarrow 2P_{3/2}(F' = 5/2)$ cooling transitions of ^{40}K and ^{6}Li, respectively (see Fig. 2.3). Interference fringes, which are produced by the parallel glass windows of the respective cell, are minimized by sending the imaging beam with a slight angle through these windows, which makes the interference period too small to be resolved.

Each of the installed imaging systems has its own CCD camera, whereas the same camera model is used for both systems (PCO imaging, ref. Pixelfly qe). The camera's CCD sensor consists of 1392×1024 pixels with a pixel size of $6.45 \times 6.45\,\mu\text{m}^2$. Due to its double shutter mode the camera has a short interframing time of $5\,\mu\text{s}$, allowing to take two subsequent pictures with this short time delay. A third picture can only be

2.8. Diagnostic tools

recorded after the first picture is read out, *i.e.* after the camera's readout time, which is 45 ms. The camera has a quantum efficiency of $\sim 45\%$ for the Li wavelength and $\sim 30\%$ for the K wavelength and a dark background level of 18 counts per pixel. More than ~ 50 photons thus need to fall on a pixel in order to be detected with a signal to noise ratio better than 2, which corresponds to a light intensity of $\sim 2.5 \times 10^{-5} \, \mu\mathrm{W/cm}^2$.

The atom clouds are imaged on the CCD chip by a single 2-inch diameter lens of focal length 6 cm for the imaging system of the MOT chamber and 7.5 cm for the imaging system of the science cell. The lenses are placed at a distance of 21 cm and 26 cm from the respective atom clouds. Their numerical apertures are $NA_{\mathrm{MC}} = 0.12$ and $NA_{\mathrm{SC}} = 0.10$, respectively. The magnification of both imaging systems is ~ 0.4.

2.8.4 Practical aspects

Using Eq. (2.7) for a reliable evaluation of the recorded images requires several conditions to be fulfilled. First, the detuning of the imaging beam needs to be appropriately chosen. Samples with a high atomic density or large atom numbers can be optically dense for weak resonant laser beams. The frequency of the imaging beam should thus be detuned in order to keep the peak optical density well below the maximum detectable value (which is ~ 6 in our system). However, for non-zero detuning the atoms refract the imaging light, which can yield false absorptive signals, when the refracted light is not collected entirely by the imaging system. In our system we indeed observe such a degradation of the image quality for a large detuning. Since the refraction angle is approximately inversely proportional to the size of the atom cloud, this effect is particularly important for small clouds. Then, the image should be taken after an appropriate time of flight, during which the cloud freely expands.

Second, the duration of the imaging pulses needs to be correctly chosen. It is a compromise between signal to noise ratio and blurring of the recorded image due to recoil-induced motion of the atoms. In addition, a too long duration might lead to a depumping of the atoms into a dark hyperfine state, requiring additional repumping light. The blurring can be estimated as follows. If an atom scatters N photons during the imaging pulse of length Δt, it gains a velocity of $N v_{\mathrm{rec}}$ along the direction of the imaging beam due to absorption and a mean velocity of $\sqrt{N} v_{\mathrm{rec}}$ in a random direction due to spontaneous emission. The resulting typical longitudinal displacement is $\Delta z = (N/2 + \sqrt{N/3}) v_{\mathrm{rec}} \Delta t$. For a proper choice of the imaging duration the Doppler shift should be negligible and the longitudinal displacement should be smaller than the depth of field of the imaging system. Also, the transverse displacement should be smaller than the resolution of the imaging system. In our setup, we chose an imaging pulse duration of 100 μs. For the used imaging beam intensity $I_{\text{img}} \sim 0.01 I_{\text{sat}}$, each atom scatters ~ 15 photons in case of resonant imaging, leading to a longitudinal displacement of $\sim 75\,\mu$m for ^{6}Li and of $\sim 10\,\mu$m for ^{40}K and a maximum Doppler shift of $\Delta\omega_D = 0.3\,\Gamma$ for ^{6}Li and $\Delta\omega_D = 0.03\,\Gamma$ for ^{40}K. For a longer duration the Doppler shift for ^{6}Li would significantly affect the absorption. The recoil-induced motion is much more limiting for ^{6}Li than for ^{40}K, due to the lighter mass of ^{6}Li.

For the chosen imaging beam pulse duration, the depumping of the atoms into a dark hyperfine state can be negligible. We verified this by comparing the optical densities in absence and presence of a repumping beam. The optical density was found larger by only 8% for ^{6}Li in the latter case. For ^{40}K, no significant change was observed due to the larger hyperfine splitting of the excited state. We thus do not use repumping light for imaging. Then, however, only those atoms can be detected which initially occupy the correct hyperfine ground state. When the trapped atoms occupy both hyperfine ground states before the imaging, it is thus necessary to optically pump all of them into the correct hyperfine state. We achieve this by exposing the atoms for 500 μs to a repumping light pulse before the image is taken. In our experimental sequence, this procedure needs to be applied only when imaging the atoms trapped in the MOT.

For a convenient quantitative analysis of the recorded transmission profiles it is desired that the imaging beam has the same intensity in both the absorption and the reference picture. Since the intensity of the imaging beam usually fluctuates, it is thus important that the time delay between the two imaging pulses is smaller than the time scale of these fluctuations, which in our setup was found to be of the order of ~ 50 ms. We recorded the dependence of the intensity difference of two subsequent imaging pulses on their time delay for three different time delays (see Fig. 2.15 (a)). The figure shows that the intensity difference can indeed be significantly decreased by choosing a small time delay. However, in the imaging sequence a small delay in general leads to an absorption of both imaging beam pulses by the atoms, since those don't have the time to disappear before the second pulse is launched. The absorption of the second pulse can, however, be circumvented by detuning the imaging beam far off resonance (by $-10\,\Gamma$). This is achieved by using an AOM in double-pass configuration, which allows to change the imaging beam frequency on a time scale of several tens of microseconds without affecting the intensity of the beam. In our system we implemented this fast imaging technique. As a result, we obtain a very reliable determination of the optical density. For example, for *a priori* identically prepared atom clouds in a (small) ^{6}Li-MOT, we obtain a standard deviation of $\sigma(\text{OD})/\overline{\text{OD}} = 3.8\%$ for the measured optical density.

2.8. Diagnostic tools

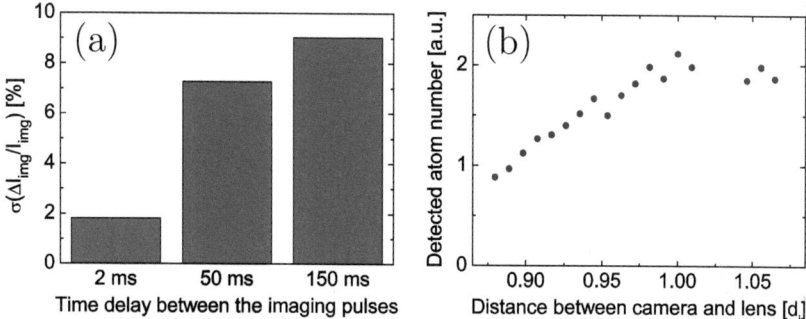

Figure 2.15: (a) Standard deviation $\sigma(\Delta I_{\text{img}}/I_{\text{img}})$ of the relative intensity difference of two subsequent imaging beam pulses for three different time delays. The intensity difference is smaller for smaller time delays. Thus the reliability of absorption images can be increased by decreasing the time delay between the two imaging pulses. (b) Detected atom number as a function of the camera position. d_i denotes the distance of the image plane from the lens. The atom number is maximum when the camera is placed in the image plane.

Finally, the imaging optics need to be correctly aligned, *i.e.* the magnification needs to be precisely known and the CCD chip needs to be placed in the focal plane of the image. We measured the magnification of our imaging systems with a "printed card" grid, which was placed at the same distance as the atom cloud. Once the magnification is known, the atom number can be reliably measured. We studied the sensitivity of the detected atom number with respect to the positioning of the CCD chip. The result is shown in Fig. 2.15 (b), which depicts the detected atom number as a function of the camera's position. The figure demonstrates that the detected atom number is maximum when the camera is placed in the image plane.

2.8.5 Auxiliary detection systems

Besides the absorption imaging systems, we installed three additional probing systems to facilitate the day-to-day operation of the experiment or to do quantitative measurements which allow a continuous measurement process. All three additional probing systems serve to measure the number of atoms trapped in the MOTs only. Since those are continuously excited by the MOT beams, they emit fluorescence light, which is proportional to the atom number. The first probing system collects a part of this fluorescence and records it with a video camera connected to a TV-screen. No quantitative measurements are done with this system. Its simply yields a convenient way to obtain approximate information about the positioning of the MOTs, their shape and their size.

The other two probing systems employ each a large area photo diode which records the power of the emitted fluorescence. Optical frequency filters allow to measure the fluorescence light of each MOT separately. The photo diodes continuously measure the atom number in the MOTs without affecting the atoms and are thus mainly employed for atom number optimization procedures. We also use them for the photoassociation experiment described in chapter 4.

The absolute atom number estimation from the measured fluorescence signals of the MOT is much less precise than the estimation from absorption images. This is, because the number of excited atoms in the MOT depends on many parameters which are difficult to determine, such as the effective detuning of the MOT beams (which depends on the magnetic field) and the intensities of cooling and repumping light. Nonetheless, when calibrated to the atom number estimates obtained from the absorption images, the fluorescence yields good estimates, since the fluorescence is in general proportional to the atom number. However, the proportionality only holds for small clouds ($N \lesssim 3 \times 10^9$ atoms) as demonstrated by Fig. 2.16, which shows the time evolution of the atom number in the ^{40}K-MOT measured by absorption images and by the fluorescence signal (calibrated to a small atom number). We attribute the deviation from proportional behavior for large atom numbers to reabsorption of the scattered photons by the atoms and to the increasing absorption of the trap light which leads to a reduced intensity in the center of the trap. Therefore we use the fluorescence light for atom number estimations only for small atom numbers.

2.8.6 Experiment control and data acquisition

We use two computers to run our experiment. One computer controls all digital, analog and GPIB commands and the cameras. The other is used for data acquisition and evaluation. Both computers run Windows XP as an operating system.

Our experiment requires precise temporal control of a variety of parameters. Most operations require a timing resolution on the millisecond scale, certain others, such as imaging, require timing on the microsecond scale. The programming software we chose is $C^\#$, which is convenient to use. Running it in a Windows environment, unfortunately brings about the disadvantage, that the operating system can interrupt the program at any time. The resulting timing of the program can thus significantly vary (typically 10 ms shot-to-shot). In order to circumvent this problem, we use an external digital input/output (DOI) board as the main clock in our system, which has an internal oscillator with 50 ns time resolution. When the experimental sequence is launched, the parameters of the sequence are loaded from the program into the buffer of this

2.8. Diagnostic tools

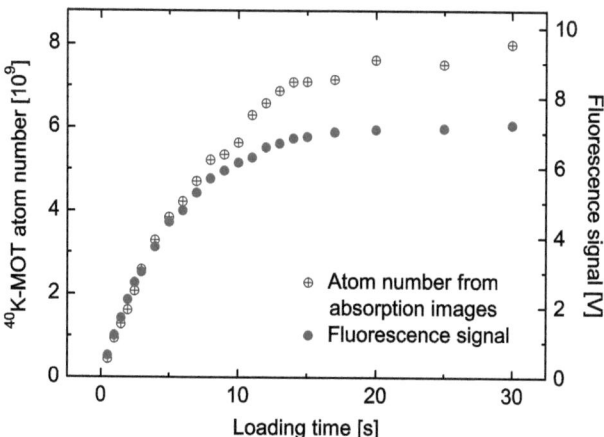

Figure 2.16: Number of trapped atoms in the ^{40}K-MOT and corresponding emitted fluorescence signal as a function of the loading time. For small atom numbers ($N \lesssim 3 \times 10^9$) the fluorescence signal is approximately proportional to the atom number. For larger atom numbers the fluorescence signal deviates from a proportional behavior. We attribute this to reabsorption of the scattered photons by the trapped atoms and to absorption of the trap light leading to a reduced intensity at the cloud center.

board and then executed. From that point on (until the buffer is cleared), the board no longer communicates with the computer and is thus not susceptible to operating system interruptions.

The loading of the parameters into the buffer is time consuming and can significantly decrease the repetition rate of the experiment. Therefore, the number of data points is kept small by maximally discretizing the time interval of the sequence. Operations which require timing on a millisecond scale are controlled with a time resolution of 0.5 ms, all others with a time resolution of 10 µs.

The DOI board controls three analog output (National Instruments, ref. NI PXI-6713) and two digital output cards (National Instruments, ref. NI PXI-6533). The analog output cards have 8 BNC outputs each, which deliver voltages between −10 V and +10 V and maximum currents of 250 mA. The digital output cards have 24 and 6 BNC outputs delivering either 0 V or 5 V. In order to avoid electronic feedback from the experiment into the computer, which might be induced by the fast switching of magnetic coils, most of the digital outputs are isolated from the experiment via optocouplers.

The data acquisition program is based on the one described in the thesis of Martin Teichmann [161]. It has been written in the programming language Python. The program instantaneously evaluates the images recorded during an imaging process. It displays the density profiles of each image and calculates and displays the normalized transmission profile. It further calculates the quantities of interest such as the total atom number, the cloud size, its central optical density, etc.

2.9 Conclusion and outlook

In this chapter we have presented the central parts of the constructed experimental Fermi mixture machine. These include the vacuum system, two laser systems for the two atomic species, two sources of high-flux cold atomic beams, a large atom number dual-species magneto-optical trap, two magnetic traps, linked by a magnetic transport, imaging devices for probing the atoms and a computer control system for the experimental procedures. To reach the quantum degenerate regime, evaporative cooling in the final magnetic trap needs to be performed. For further investigation, the atom clouds will be transfered into an optical dipole trap. To conclude this chapter we show two pictures of our laboratory recorded at the beginning and end of the construction work (see Fig. 2.17).

2.9. Conclusion and outlook 59

Figure 2.17: Photographs of the laboratory recorded at the beginning (up) and end (down) of the construction work.

Chapter 3

Characterization of the experimental apparatus

In this chapter we present the characterization of the constructed experimental setup. We start with a detailed study of the atom sources, *i.e.* the Zeeman slower for ^{6}Li and the 2D-MOT for ^{40}K. Then we characterize the dual-species MOT, first in single-species and subsequently in dual-species operation for which we studied interspecies collisions. Finally, we describe the implementation of the subsequent stages in the experimental sequence for the creation of a degenerate mixture, including the transfer of the atomic clouds into the magnetic quadrupole trap, the magnetic trapping and transport to the final science cell.

3.1 ^{6}Li Zeeman slower

For our application the essential parameter which characterizes the performance of the Zeeman slower is the capture rate of the ^{6}Li-MOT. We studied its dependence as a function of several Zeeman slower parameters, such as the temperature of the oven, the power of the slowing light, the magnitude of the magnetic field and the intensity ratios between the repumping and slowing light. The optimized values of these parameters are displayed in Tab. 2.1, leading to a ^{6}Li-MOT capture rate of $\sim 1.2 \times 10^9$ atoms/s.

Figure 3.1 (a) shows the dependence of the ^{6}Li-MOT capture rate on the power of the Zeeman slowing light. The curve increases with increasing beam power and indicates saturation for higher powers. In the experiment the slowing light power is 45 mW, for which the curve in Fig. 3.1 (a) starts to saturate, demonstrating that, for the given beam power, the size of the slowing beam is well chosen. In particular it shows that the beam is not absorbed significantly by the atoms inside the slower.

The dependence of the ^{6}Li-MOT capture rate on the intensity ratio between re-

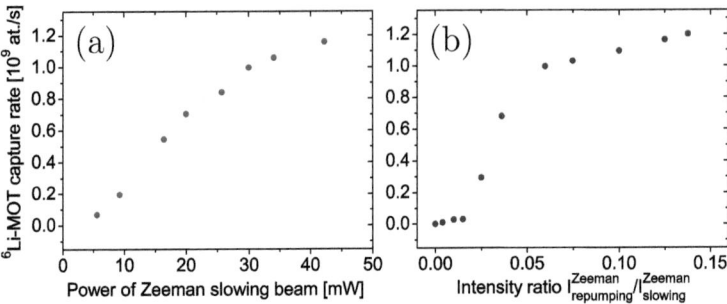

Figure 3.1: ^{6}Li-MOT capture rate as a function of (a) the power of the Zeeman slowing light for a constant repumping light power of 5.6 mW and (b) the intensity ratio between repumping and slowing light of the Zeeman slower for a constant slowing light power of 45 mW. The intensities of the superimposed beams depend on the position inside the slower, since the beams are focused toward the oven. At the position where the magnetic field changes sign, a power of 10 mW corresponds to an intensity of 2.5 I_{sat}, with the saturation intensity I_{sat} given in Tab. 2.3.

pumping and slowing light of the Zeeman slower is depicted in Fig. 3.1 (b). The curve increases with increasing repumping intensity and saturates for higher intensities. For the intensity ratio $I_{\text{rep}}/I_{\text{slow}} \sim 0.1$ the repumping intensity in the region where the magnetic field of the Zeeman slower changes sign, is of the order of the saturation intensity. Therefore the transition probability of the repumping transition saturates at $I_{\text{rep}}/I_{\text{slow}} \sim 0.1$, explaining the behavior in Fig. 3.1 (b). The graph shows that the Zeeman slower only requires a small repumping intensity. It is important that the repumping light has the same circular polarization as the slowing light, since it helps to optically pump the atoms to the cycling transition used for slowing.

Figure 3.2 (a) shows the ^{6}Li-MOT capture rate as a function of the magnitude of the axial magnetic field of the Zeeman slower. The position of the maximum depends on the detuning of the slowing light, which, in the experiment, was chosen to be $\Delta\omega_{\text{slow}} = -75\,\Gamma$. The experimentally determined optimum value of the magnetic field is close to the value expected from our numerical simulations. For values which differ from the optimum one, the profile of the axial magnetic field is less well adapted to the change in the atoms' velocity, leading to a less efficient slowing.

Figure 3.2 (b) shows the dependence of the ^{6}Li-MOT capture rate on the oven temperature T (circles) as well as a (scaled) theoretical prediction (solid curve) for the experimental data. The curve shows a nearly exponential increase of the capture rate with the temperature. The theoretical prediction is based on a model which assumes no collisions between the atoms (i.e., no intrabeam collisions and no collisions between

3.1. ^{6}Li Zeeman slower

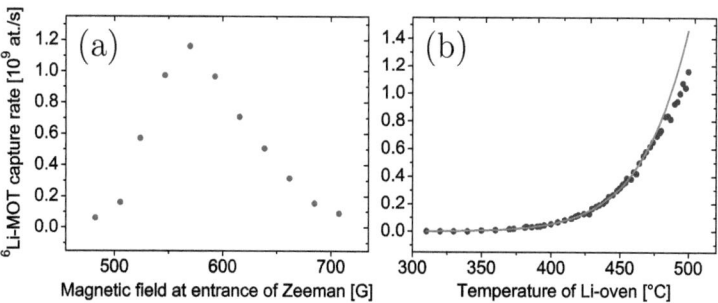

Figure 3.2: ^{6}Li-MOT capture rate as a function of (a) the axial magnetic field of the Zeeman slower and (b) the temperature of the Li-oven. Circles represent the experimental data and the solid curve the theoretical prediction from Eq. (3.3).

the beam and the MOT atoms). It is derived as follows.

In the absence of collisions, the normalized velocity distribution of the Zeeman-slowed atoms exiting the slower does not depend on the temperature of the oven. Assuming that the ^{6}Li-MOT captures mainly atoms which have been slowed by the Zeeman slower, the capture rate $\dot{N}_\text{M}$ of the ^{6}Li-MOT is a temperature-independent fraction of the flux $\dot{N}_\text{Z}$ of the Zeeman-slowed atoms: $\dot{N}_\text{M}(T) = \kappa_1 \dot{N}_\text{Z}(T)$. The proportionality constant κ_1 depends on the divergence of the atomic beam and the capture velocity of the ^{6}Li-MOT. The flux of the Zeeman-slowed atoms $\dot{N}_\text{Z}$ is given by the flux of the oven atoms which have a speed smaller than the Zeeman slower's capture velocity v_cap^Zee and which are in the correct internal atomic state to be decelerated by the Zeeman slower (i.e. $F = 3/2$, $m_F = 3/2$). Assuming the oven to be in thermal equilibrium, $\dot{N}_\text{Z}$ is given by [138, 144]

$$\dot{N}_\text{Z}(T) = \kappa_2 n_\text{s}(T) A \int_0^{\Omega_\text{Z}} d\Omega \frac{\cos\theta}{4\pi} \int_0^{v_\text{cap}^\text{Zee}} v f(v,T) dv, \quad (3.1)$$

with a temperature-independent constant κ_2, which equals the fraction of atoms which are in the correct internal atomic state. $n_\text{s}(T)$ is the atomic density in the oven, $A = 2 \times 10^{-5}\,\text{m}^2$ the aperture surface of the oven, $\Omega_\text{Z} = A'/l^2 = 5 \times 10^{-4}$ the solid angle of the atomic beam (with A' the aperture surface of the last differential pumping tube and l the distance between the two aperture surfaces A, A') and $d\Omega = 2\pi \sin\theta d\theta$, with θ the emission angle with respect to the oven axis. $f(v, T)$ is the normalized speed distribution function given by

$$f(v,T) = \sqrt{\frac{2m^3}{\pi k_B^3 T^3}} v^2 \exp\left(-\frac{mv^2}{2k_B T}\right). \tag{3.2}$$

Since the solid angle of the atomic beam is small, it is $\cos\theta \approx 1$ and thus $\int_0^{\Omega_z} d\Omega \cos\theta \approx \Omega_z$.

The explicit temperature dependence of the ^{6}Li-MOT capture rate is then obtained via $\dot{N}_M(T) = \kappa_1 \dot{N}_Z(T)$ by substituting into Eq. (3.1) the ideal gas equation $n_s(T) = p_s/(k_B T)$ and the relation $p_s = p_a \exp[-L_0/(k_B T)]$ for the saturated vapor pressure p_s, with $p_a = 1.15 \times 10^8$ mbar and the latent heat of vaporization $L_0/k_B = 18474$ K [162]. This relation applies to the temperature range 300-500 °C with an accuracy of 5%. Thus, we have

$$\dot{N}_M(T) = \kappa A \Omega_z p_a \sqrt{\frac{m^3}{8\pi^3 k_B^5 T^5}} e^{-\frac{L_0}{k_B T}} \int_0^{v_{\text{cap}}^{\text{Zee}}} v^3 e^{-\frac{mv^2}{2k_B T}} dv, \tag{3.3}$$

with $\kappa = \kappa_1 \kappa_2$. Scaling Eq. 3.3 to the experimental data for a given (low) temperature ($T = 350$°C) yields the theoretical prediction for the curve shown in Fig. 3.2. The scaling yields $\kappa = 10^{-3}$, thus 0.1% of the atoms, which enter the Zeeman slower with a velocity smaller than $v_{\text{cap}}^{\text{Zee}}$, are captured by the ^{6}Li-MOT.

The main contribution to the small value of κ is the large divergence of the slowed atomic beam: κ is proportional to the ratio of the atomic beam cross section and the capture surface of the ^{6}Li-MOT, which is estimated to $\sim 10^{-2}$ (assuming the ^{6}Li-MOT capture surface to be a circle of 1.1 cm diameter). Two-dimensional transverse laser cooling of the atomic beam could vastly increase the value of κ. Still, not all atoms which enter the Zeeman slower with a velocity smaller than $v_{\text{cap}}^{\text{Zee}}$ and which reach the capture surface of the ^{6}Li-MOT will be trapped. The value of κ implies that only 10% of those atoms will be trapped. This is due to an inefficient capture of the ^{6}Li MOT and to a significant fraction of oven atoms occupying the incorrect internal atomic states.

The obtained theoretical prediction agrees well with the experimental data for temperatures below 475 °C (see Fig. 3.2 (b)). For temperatures above 475 °C, the experimental data deviate from the prediction indicating that intrabeam collisions or collisions between the atoms in the beam and the MOT become important. We found that for T=500 °C collisions between the thermal ^{6}Li beam and the trapped ^{6}Li-MOT atoms indeed take place, which we verified by measuring the lifetime of the ^{6}Li-MOT in presence and absence of the thermal ^{6}Li beam, making use of the mechanical block placed at the exit of the oven. The lifetime was found 10% larger for the case where the thermal ^{6}Li beam was blocked. In a similar way the thermal ^{6}Li beam also affects the lifetime of the ^{40}K-MOT. In order to avoid a reduction of the number of trapped ^{40}K atoms in the dual-species MOT, we therefore limit the ^{6}Li-oven temperature to 500 °C.

With the help of Eq. 3.1 the lifetime of the oven can be estimated. Assuming that the collimation tube of the oven recycles all atoms sticking to its wall and the vacuum pumps have no impact on the Li pressure in the oven, the total atomic flux through the collimation tube is obtained by replacing $A' = A$, $v_{\text{cap}}^{\text{Zee}} = \infty$ and $l =$ 8 cm (the length of the collimation tube) in Eq. 3.1. For the working temperature $T = 500$ °C the lithium vapor pressure is $p_s = 4.8 \times 10^{-3}$ mbar, corresponding to a density $n_s = 4.5 \times 10^{19}$ m^{-3}. Thus, the atom flux through the collimation tube is $\dot{N}_\text{O} = 3.5 \times 10^{14}$ s$^{-1} \hat{=} 3.5 \times 10^{-12}$ kg/s. With 3 g of ^{6}Li this corresponds to an oven lifetime of $\tau_{\text{oven}} \sim 25$ years. (The importance of the recycling becomes manifest when comparing this value to the hypothetical lifetime of the oven, would the collimation tube be replaced by an aperture of the same surface. In this case the atom flux through this aperture would be $\dot{N}_\text{O}^{\text{hyp}} = (\pi l^2 / A) \dot{N}_\text{O} \sim 1000 \dot{N}_\text{O}$ and thus $\tau_{\text{oven}}^{\text{hyp}} \sim 10$ days.)

3.2 ^{40}K 2D-MOT

For our purpose the essential parameter which characterizes the performance of the 2D-MOT is the capture rate of the ^{40}K-MOT. We studied its dependence as a function of several 2D-MOT parameters, such as: the vapor pressure in the 2D-MOT cell, the total cooling light power, the detuning of the cooling frequency and the intensity ratios between the repumping and cooling light and between the pushing and retarding beams. Further, we determined the mean velocity of the atoms in the atomic beam. The optimized values of the essential parameters of the ^{40}K 2D-MOT are displayed in Tab. 2.2, leading to a ^{40}K-MOT capture rate of $\sim 1.4 \times 10^9$ atoms/s.

The mean velocity of the atoms in the atomic beam can be experimentally estimated as follows. Due to the Doppler shift between the atoms in the beam and the pushing beam, the atoms can be considered to perform a ballistic flight when leaving the 2D-MOT region. Thus, their mean velocity is approximately given by the average time required for the atoms to reach the 3D-MOT. This time was measured by recording the time delay of the onset of the ^{40}K-MOT loading after switching on the 2D-MOT beams. The time delay is obtained from the loading curve of the ^{40}K-MOT, which is depicted in Fig. 3.3, and a linear fit of the data points. The intersection of the fit with the time-axis yields the desired time delay, which is ~ 23 ms. The distance between the centers of the 2D- and 3D-MOT being 55 cm, we deduce a mean longitudinal velocity of the captured atoms of ∼24 m/s. At this velocity, the displacement due to gravity of the beam of atoms from the ^{40}K-MOT center is ∼2.6 mm, which is negligible compared to the size of the ^{40}K-MOT beams and the divergence of the atomic beam. The measured velocity of the captured atoms is a lower bound of the capture velocity of the ^{40}K-MOT: $v_{\text{cap}}^{\text{KMOT}} > 24$ m/s. Furthermore, Fig 3.3 allows us to deduce the loading rate of the ^{40}K-MOT, which is given by the slope of the fitted line and amounts to ∼ 1.4×10^9 atoms/s.

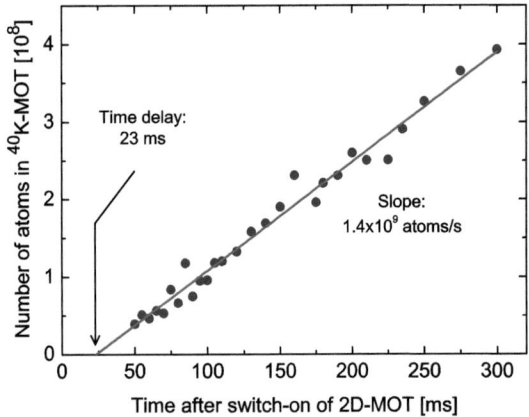

Figure 3.3: Number of atoms loaded in ^{40}K-MOT as a function of the time after switch-on of the 2D-MOT. The onset of the loading has a time delay of ∼ 23 ms, which determines the mean longitudinal velocity of the captured atomic beam atoms to ∼24 m/s. This velocity further represents a lower bound for the capture velocity of the ^{40}K-MOT: $v_{\text{cap}}^{\text{KMOT}} > 24$ m/s. The slope of the curve yields the capture rate of the ^{40}K-MOT, which is ∼ 1.4×10^9 atoms/s.

Figure 3.4 (a) shows the dependence of the ^{40}K-MOT capture rate on the detuning $\Delta\omega_{\text{cool}}$ of the 2D-MOT cooling light. The curve has a maximum at $\Delta\omega_{\text{cool}} = -3.5\,\Gamma$ and a full width at half maximum (FWHM) of $2.7\,\Gamma$. The maximum is the result of two opposing effects: the scattering force of the 2D-MOT beams decreases with increasing detuning whereas the capture velocity increases [135]. The first effect implies a less efficient transverse cooling whereas the second leads to a more efficient capture of atoms. An additional effect might influence the shape of the curve: since the scattering force of the pushing beam depends on the detuning, also the mean-velocity of the atomic beam depends on it [140, 141, 143]. Since we measure the ^{40}K-MOT capture rate rather than the flux of the 2D-MOT, the mean-velocity might exceed the capture velocity of the

3.2. ^{40}K 2D-MOT

^{40}K-MOT. However, as shown in Refs. [140, 141, 143], the mean-velocity of the beam only slightly changes with the detuning, such that we expect this effect to only weakly influence the curve. In summary, the shape of the curve in Fig. 3.4 (a) shows that the ^{40}K-MOT capture rate is not very sensitive to changes of $\Delta\omega_{\text{cool}}$.

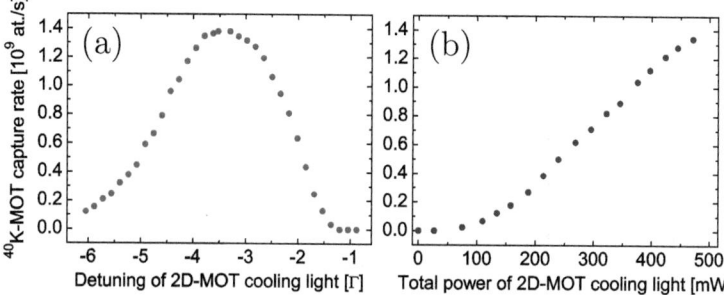

Figure 3.4: ^{40}K-MOT capture rate as a function of (a) the detuning and (b) the total power of the cooling light used for the 2D-MOT (for a constant intensity ratio between the cooling and repumping light). The total power refers to the sum of the powers in the six 2D-MOT beams, where a power of 470 mW corresponds to a total intensity of $\sim 47\, I_{\text{sat}}$ at the center of the 2D-MOT, with the saturation intensity I_{sat} given in Tab. 2.3.

The dependence of the ^{40}K-MOT capture rate on the total power of the 2D-MOT cooling light is depicted in Fig. 3.4 (b). The total power refers to the sum of the powers in the six 2D-MOT beams. According to the chosen beam sizes, the maximum power of 470 mW corresponds to a total intensity of $\sim 47\, I_{\text{sat}}$ (for zero detuning) at the center of the 2D-MOT, with the saturation intensity I_{sat} given in Tab. 2.3. The curve almost linearly increases with light power without a clear indication of saturation. The increase is due to two effects. First, the 2D-MOT capture velocity increases with laser power due to the power broadening of the atomic spectral lines. Second, the scattering force increases, resulting in a steeper transverse confinement, which facilitates the injection of the atoms into the differential pumping tube. At some point, the curve is expected to saturate. This is, because on the one hand, the optical transition used for cooling will saturate, and on the other hand, the temperature of the cooled atoms and light-induced collisions between them increase with light power. The latter effects, however, are less limiting in a 2D-MOT as compared to a 3D-MOT, since the atomic density in a 2D-MOT is typically three orders of magnitude smaller due to the absence of a three-dimensional confinement. Thus, in a 2D-MOT a high light power would be required to reach the regime of saturation.

Figure 3.5 (a) shows the dependence of the ^{40}K-MOT capture rate on the intensity ratio between the cooling and repumping light of the 2D-MOT for the two different repumping detunings $\Delta\omega_{\text{rep}}^{(1)} = -2.5\,\Gamma$ and $\Delta\omega_{\text{rep}}^{(2)} = -6.5\,\Gamma$ and for a constant total cooling light power of 300 mW. The graph shows that for both frequencies the ^{40}K-MOT capture rate increases with increasing repumping intensity and that it saturates at high intensities. It also shows that the maximum capture rate is bigger for the smaller detuning. The intensity dependence of the curves results from the likewise intensity dependence of the transition probability for an atomic transition. The maximum capture rate is bigger for the smaller detuning, since this detuning contributes more efficiently to the cooling process. In our experiment, a fixed total laser power is available for both repumping and cooling light. It is distributed such that the resulting capture rate is maximized. It was found to be maximum for an intensity ratio of $I_{\text{rep}}/I_{\text{cool}} \sim 1/2$. For that ratio the detuning $\Delta\omega_{\text{rep}}^{(2)} = -2.5\,\Gamma$ also yields the maximum capture rate.

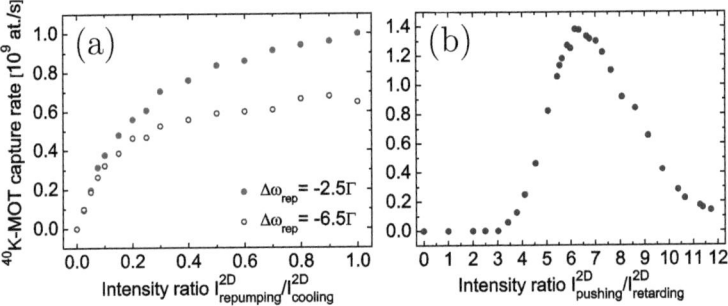

Figure 3.5: ^{40}K-MOT capture rate as a function of the intensity ratio between (a) repumping and cooling light of the 2D-MOT for two different repumping detunings $\Delta\omega_{\text{rep}}$ and a constant total cooling light power of 300 mW (which corresponds to a total intensity of $\sim 30\,I_{\text{sat}}$) and (b) the pushing and the retarding beams of the 2D-MOT. The intensities of the pushing and retarding beams refer to the intensities along the atomic beam axis.

The dependence of the ^{40}K-MOT capture rate on the intensity ratio between pushing and retarding beam is depicted in Fig. 3.5 (b). The curve has a maximum at $I_{\text{push}}/I_{\text{retard}} \sim 6$. It is zero for values of $I_{\text{push}}/I_{\text{retard}}$ between 0 and 3, then increases until the maximum and falls off again with a smaller slope. This feature is a consequence of the reflectivity of the mirror inside the vacuum and of the size of its hole. For a given intensity ratio $I_{\text{push}}/I_{\text{retard}}$ along the (horizontal) direction of the atomic beam, the mirror's reflectivity determines the intensity ratio $I_{\text{push}}^*/I_{\text{retard}}^*$ along the vertical direction above the reflecting surface of the mirror (see Fig. 2.9). If $I_{\text{push}}^*/I_{\text{retard}}^*$ differs

3.2. ^{40}K 2D-MOT

from 1, the atomic beam can experience a vertical deflection in this region. The hole inside the mirror creates a dark cylinder in the pushing beam after its reflection, so that in the region above the hole only light from the retarding beam has a vertical direction, which can also give rise to a vertical deflection of the atomic beam.

In the following we estimate the deflection of the atomic beam, which is induced by the unbalanced retarding beam in the small region above the hole. Assuming the atomic beam to have reached its final longitudinal velocity of 24 m/s when entering into the hole, the atoms spend 85 µs in the region above the hole. Neglecting Doppler shifts and the presence of the pushing beam along the horizontal direction (no transverse beams are present in the region above the mirror), the atoms will scatter $N_{\text{ph}} = R_{\text{sc}} \times (85\,\mu s) \sim$ 75 photons, with R_{sc} being the scattering rate [135] for the given detuning $\Delta\omega_{\text{cool}} = -3.5\,\Gamma$ and peak intensity $I^*_{\text{retard}} = 2.5 I_{\text{sat}}$. The recoil velocity of ^{40}K being given by $v_{\text{rec}} = 0.013$ m/s, each atom will accumulate a transverse velocity of $v_{\text{dev}} \sim 1$ m/s. This leads to a downwards deflection of the atomic beam by an angle of ~ 40 mrad, which is more than a factor two bigger than the maximum deflection angle allowed by the differential pumping tubes. The atoms will thus not reach the ^{40}K-MOT.

This deflection needs to be compensated by an intensity imbalance $I^*_{\text{push}} > I^*_{\text{retard}}$ in the region above the reflecting surface of the mirror, as that results in an upwards deflection of the atomic beam. For the given mirror reflectivity of 50%, $I^*_{\text{push}} > I^*_{\text{retard}}$ is equivalent to $I_{\text{push}}/I_{\text{retard}} > 4$, which corresponds to the experimental observation depicted in Fig. 3.5 (b). The deflection of the atomic beam in the region above the hole could be avoided using a beam block inside the retarding beam which creates a shadow on the mirror hole. In this configuration the position of the curve optimum in Fig. 3.5 (b) would change from $I_{\text{push}}/I_{\text{retard}} = 6$ to $I_{\text{push}}/I_{\text{retard}} = 4$. For mirrors with a reflectivity close to 100% the position of the curve optimum could thus even be changed to $I_{\text{push}}/I_{\text{retard}} = 1$, for which the longitudinal optical molasses cooling would be most efficient leading to a maximum 2D-MOT flux. When such a perfect reflectivity is not available, this optimum ratio could still be obtained, that is, by using two beam blocks (one per beam), which create shadows on the mirror in the region where it otherwise reflects parts of the pushing beam onto the atomic beam. The dark region inside the longitudinal beams, might then, however, restrict the longitudinal cooling efficiency. We did not implement these configurations since the longitudinal optical molasses cooling is still very efficient even for the given intensity imbalance of 6 along the atomic beam axis, due to the polarization gradients generated by the transverse 2D-MOT beams, such that the implementation of the described ideal configurations might not be very gainful.

We now study the dependence of the ^{40}K-MOT capture rate on the vapor pressure of potassium (all isotopes) in the 2D-MOT cell, which is shown in Fig. 3.6 (circles) together with a fit to a theoretical model (solid curve). The vapor pressure was determined from the absorption profile of a low intensity probe according to the procedure described in Appendix A. The curve in Fig. 3.6 has a maximum at a vapor pressure of 2.3×10^{-7} mbar. In the absence of collisions, the curve should increase linearly with pressure, which is indeed observed for low pressures. For high pressures, collisions become important and limit the ^{40}K-MOT capture rate. The dependence of the ^{40}K-MOT capture rate L on the pressure p can be described by the function [143]

$$L = L_0 \exp\left[-\left(\Gamma_{\text{coll}} + \beta \int n^2(\mathbf{r}) d^3 r\right) \langle t_{\text{cool}} \rangle\right], \tag{3.4}$$

where L_0 denotes the hypothetical capture rate of the ^{40}K-MOT in the absence of collisions in the 2D-MOT chamber, Γ_{coll} denotes the collisional loss rate due to collisions in the 2D-MOT chamber between the cooled atoms and the background atoms, $\langle t_{\text{cool}} \rangle$ is the average time which the atoms spend inside the 2D-MOT cooling region, $n(\mathbf{r})$ is the position-dependent atomic density in the atomic beam, and β is the two-body loss rate coefficient which describes the cold collisions between the ^{40}K atoms in the atomic beam. L_0 is proportional to the atomic density n_K in the vapor cell, and $\Gamma_{\text{coll}} = n_K \sigma_{\text{eff}} \langle v \rangle$, where σ_{eff} is the effective collision cross section, and $\langle v \rangle \sim 400$ m/s the mean velocity of the thermal potassium atoms. The term describing the cold collisions is approximately proportional to n_K^2 due to the small density obtained in the 2D-MOT. For the investigated pressure range, the ratio p/n_K only changes slightly with temperature and can thus be considered constant. Therefore Eq.(3.4) can be written as

$$L(p) = \kappa_1 p \exp\left(-\kappa_2 p - \kappa_3 p^2\right), \tag{3.5}$$

with the constants $\kappa_1, \kappa_2, \kappa_3$, which are obtained from the fit shown in Fig. 3.6. At the curve's maximum, the fit yields $\kappa_2 p / \kappa_3 p^2 = 8$, showing that the collisions which limit the ^{40}K-MOT capture rate are mainly the collisions with the hot background atoms, consisting mostly of ^{39}K.

The background atoms are predominantly potassium atoms. These can collide either with the excited or the non-excited ^{40}K atoms of the atomic beam. Depending on the isotopes of the colliding partners, these collisions have different cross sections. Collisions between an excited and a non-excited atom of the same isotope usually have a very large cross section due to the strong long-range resonant dipole interaction, described by a C_3/R^3-potential. In 2D-MOT systems of other atomic species these

3.2. ^{40}K 2D-MOT

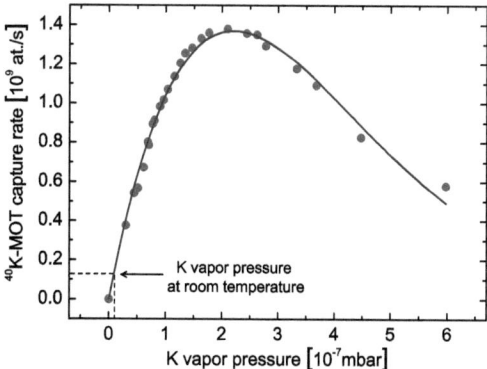

Figure 3.6: ^{40}K-MOT capture rate as a function of the potassium vapor pressure (all isotopes). Circles: experimental data, solid curve: fit of the experimental data by Eq. (3.5). Due to the low abundance of the ^{40}K-isotope in our potassium sample (4%), the ^{40}K-MOT capture rate is limited by collisions between the ^{40}K atoms and the other K-isotopes in the 2D-MOT cell. At room temperature the potassium vapor pressure is 1×10^{-8} mbar.

collisions have been identified as the ones which limit the flux of the 2D-MOT [140, 141, 143]. In the case of ^{40}K, the scattering rate for these collisions is reduced by the small abundance of ^{40}K in the vapor. Therefore other collisions might limit the flux. In order to identify the flux-limiting collisions we calculate the cross section of different possible collisions and deduce the corresponding collision rates. The cross sections can be calculated using the approach described in Ref. [163] for losses out of a cold atom cloud. The cross section for collisions involving an excited and a non-excited ^{40}K atom is given by [163]

$$\sigma_{\text{eff}}^{40,40^*} = \pi \left(\frac{4C_3}{m v_{\text{esc}} \langle v \rangle} \right)^{2/3}, \tag{3.6}$$

where m is the mass of the ^{40}K atom, $v_{\text{esc}} \sim 1\,\text{m/s}$ is the estimated transverse velocity kick needed to make an atom miss the ^{40}K-MOT, and $C_3 = 5.4 \times 10^{-48}\,\text{Jm}^3$ is the dispersion coefficient for the resonant dipole-dipole interaction [164]. The cross section for collisions involving a non-excited ^{40}K atom and a non-excited K atom of the different isotopes is given by [163] (for a derivation, see also Sec. 3.5)

$$\sigma_{\text{eff}}^{40,39} \sim \sigma_{\text{eff}}^{40,41} \sim \sigma_{\text{eff}}^{40,40} = \pi \left(\frac{15\pi C_6}{8 m v_{\text{esc}} \langle v \rangle} \right)^{1/3}, \tag{3.7}$$

where $C_6 = 3.7 \times 10^{-76}\,\text{Jm}^6$ is the dispersion coefficient for the underlying van der Waals interaction [164]. Substituting the experimental parameters, one obtains: $\sigma_{\text{eff}}^{40,40*} = 2.7 \times 10^{-16}\,\text{m}^2$ and $\sigma_{\text{eff}}^{40,39} \sim \sigma_{\text{eff}}^{40,41} \sim \sigma_{\text{eff}}^{40,40} = 1.3 \times 10^{-17}\,\text{m}^2$. The resulting collision rates are proportional to the atomic densities n_{39}, n_{40} and n_{41} of the corresponding isotopes in the vapor and the relative number of excited ^{40}K atoms in the atomic beam, which was estimated to $P \sim 0.1$ for the given beam detunings and intensities. One obtains

$$\Gamma_{\text{coll}}^{40,40*} = P n_{40} \sigma_{\text{eff}}^{40,40*} \langle v \rangle = 4.4 \times 10^{-16} n_\text{K}, \quad (3.8)$$

$$\Gamma_{\text{coll}}^{40,39} = (1-P) n_{39} \sigma_{\text{eff}}^{40,39} \langle v \rangle = 4.4 \times 10^{-15} n_\text{K}, \quad (3.9)$$

$$\Gamma_{\text{coll}}^{40,40} = (1-P) n_{40} \sigma_{\text{eff}}^{40,40} \langle v \rangle = 2.0 \times 10^{-16} n_\text{K}, \quad (3.10)$$

$$\Gamma_{\text{coll}}^{40,41} = (1-P) n_{41} \sigma_{\text{eff}}^{40,41} \langle v \rangle = 3.0 \times 10^{-16} n_\text{K} \quad (3.11)$$

(n_K denoting the atomic density of potassium in the vapor cell). The dominant collision rate here is $\Gamma_{\text{coll}}^{40,39}$ (Eq. (3.9)) for collisions involving a non-excited ^{40}K atom and a non-excited ^{39}K atom from the background. The largest collision rate for collisions between two ^{40}K atoms, $\Gamma_{\text{coll}}^{40,40*}$, is by a factor of 10 smaller than $\Gamma_{\text{coll}}^{40,39}$. Therefore, collisions involving two ^{40}K atoms are not the collisions which limit the flux of the 2D-MOT. This is in contrast to 2D-MOT systems of other species. From the difference between $\Gamma_{\text{coll}}^{40,40*}$ and $\Gamma_{\text{coll}}^{40,39}$ we conclude that the flux of the 2D-MOT for ^{40}K could still be improved by about a factor of 10 by using a potassium sample of a higher isotopic enrichment.

3.3 ^{6}Li-^{40}K dual-species MOT

In this section we characterize the ^{6}Li-^{40}K dual-species MOT. We first present the characterization of the MOTs in single-species operation and then turn to the characterization of the MOT in dual-species operation. The optimum parameters, which lead to atom numbers of $N_{\text{single}} \sim 8.9 \times 10^9$ in the ^{40}K-MOT and $N_{\text{single}} \sim 5.4 \times 10^9$ in the ^{6}Li-MOT, are displayed in Tab. 2.3 together with the characteristics of the MOTs (in dual-species operation, the atom numbers only slightly change due to the additional interspecies collisions to $N_{\text{dual}} \sim 8.0 \times 10^9$ in the ^{40}K-MOT and $N_{\text{dual}} \sim 5.2 \times 10^9$ in the ^{6}Li-MOT). The $(1 - 1/e)$-loading times of the MOTs are $\sim 5\,\text{s}$ for ^{40}K and $\sim 6\,\text{s}$ for ^{6}Li.

3.3.1 Single-species MOTs

In this section we describe the characterization of the single-species magneto-optical traps using the parameters for the optimized dual-species operation. We determined the atom numbers, the atomic densities in the cloud center, the loading times and the temperatures. Furthermore, we studied for each atomic species the dependence of the steady-state MOT atom number on the following parameters: the magnetic field gradient, the power and detuning of the cooling light and the intensity ratio between the repumping and cooling light.

Magneto-optical traps with large atom numbers have a high optical density and are optically dense for weak resonant laser beams. Therefore, when determining the atom number via absorption imaging, the frequency of the imaging beam has to be detuned, so not to "black out" the image.

Figures 3.7 (a,b) depict the *detected* atom number of the two MOTs (circles) as a function of the detuning of the imaging beam. The detected atom number was derived from the measured optical density assuming the imaging beam to be resonant. The curves are expected to have the shape of a Lorentzian with the peak centered around zero detuning. The experimental data shown in Figures 3.7 (a,b) clearly deviate from a Lorentzian behavior—they saturate for small magnitudes of the detuning. This deviation demonstrates that the MOTs are optically dense in this regime. A correct estimate of the atom number is obtained from an extrapolation of the experimental data to zero detuning based on a Lorentzian fit of the curve wings (solid curves). A reliable extrapolation, however, requires imposing the width of the Lorentzian fit. In order to determine this width, an additional experiment was done (not shown): the data in Figs. 3.7 (a,b) were again recorded and fitted by a Lorentzian for a MOT with a small atom number and a low optical density (obtained by a short loading of 250 ms). The widths found by this additional measurement were $1.05\,\Gamma$ for ^{40}K and $1.5\,\Gamma$ for ^{6}Li. For ^{40}K this width corresponds to the natural linewidth of the exited state addressed by the imaging transition. For ^{6}Li the width is larger than the natural linewidth, since the small excited hyperfine structure is unresolved and thus its width ($\sim 0.5\,\Gamma$) and the natural linewidth add up (this line broadening does not occur when a bias magnetic field is applied and a closed transition is used for imaging). The peak values of the Lorentzian fits in Figs. 3.7 (a,b) finally yield the atom numbers in the MOTs, given in Tab. (2.3).

Figures 3.7 (c,d) show images of the MOTs and their doubly-integrated density profiles $\bar{\bar{n}}$ for the case of a resonant imaging beam. The flat top of $\bar{\bar{n}}$ as a function of position shows that the MOTs are optically dense. Their central optical densities for the resonant imaging beam are determined to be ~ 20 for ^{40}K and ~ 15 for ^{6}Li by the

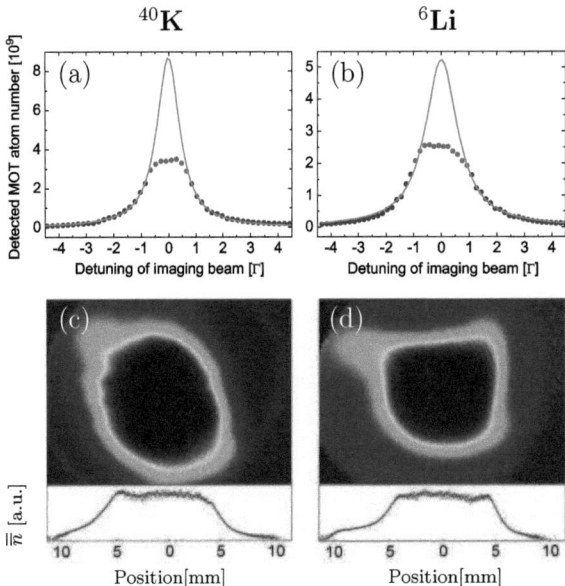

Figure 3.7: (a,b) Detected atom number in the MOTs as a function of the detuning of the imaging beams. Circles correspond to the experimental data and solid curves to Lorentzian fits of the curve wings with an imposed width, which was determined by another measurement. (c,d) Absorption images of the MOTs and the doubly-integrated density profile $\bar{\bar{n}}$, recorded with a resonant imaging beam. The graphs (a,c) relate to the ^{40}K-MOT and (b,d) to the ^{6}Li-MOT. The flat top of $\bar{\bar{n}}$ in the graphs (c,d) and the saturation of the detected atom number for small magnitudes of the detuning in the graphs (a,b) demonstrate that the MOTs are optically dense for the imaging beam when the detuning is small. Their (extrapolated) central optical densities for a resonant imaging beam are ~ 20 for ^{40}K and ~ 15 for ^{6}Li.

extrapolation technique described above. In addition, the density profiles in Figs. 3.7 (c,d) show that the MOTs have spatial extensions of the order of 1 cm.

Due to the high optical density of the MOTs all images are taken with a detuning of $-2\,\Gamma$. The reference picture (see Sec. 2.8.1) is taken with a detuning of $-10\,\Gamma$. The light absorption of the reference picture due to the presence of the atoms is thus ~ 25 times less than in the absorption picture and can be neglected.

The atomic density in the MOT center is extracted from the recorded two-dimensional density profile as described in Sec. 2.8.2. We obtain $n_c^K \sim 3 \times 10^{10}$ atoms/cm^3 and $n_c^{Li} \sim 2 \times 10^{10}$ atoms/cm^3, respectively. The temperature of the MOTs in single-species operation is determined by the time-of-flight method described

3.3. ^{6}Li-^{40}K dual-species MOT

in Sec. 2.8.2. The ^{40}K-MOT has a temperature of $290\,\mu$K and the ^{6}Li-MOT of $1.4\,$mK. Both temperatures are higher than the Doppler cooling limit, because of the high intensity in the MOT beams. In addition, for ^{6}Li, the unresolved excited hyperfine structure (see Fig. 2.3) inhibits sub-Doppler cooling effects. The same temperatures are found in dual-species operation. The measured temperatures and atomic densities yield the peak phase space densities $D_\text{K} = n_\text{c}^\text{K} \Lambda_\text{K}^3 \sim 1.2 \times 10^{-7}$ and $D_\text{Li} = n_\text{c}^\text{Li} \Lambda_\text{Li}^3 \sim 1.3 \times 10^{-7}$ with the thermal de Broglie wavelength $\Lambda = \sqrt{2\pi\hbar^2/(mk_\text{B}T)}$, respectively.

The dependence of the MOT atom number on the detuning of the cooling light is depicted in Figs. 3.8 (a,b). The atom number is maximum at $\Delta\omega_\text{cool}^\text{K} = -3\,\Gamma$ for ^{40}K and at $\Delta\omega_\text{cool}^\text{Li} = -5\,\Gamma$ for ^{6}Li, and has a FWHM of $2.3\,\Gamma$ and $4.1\,\Gamma$, respectively.

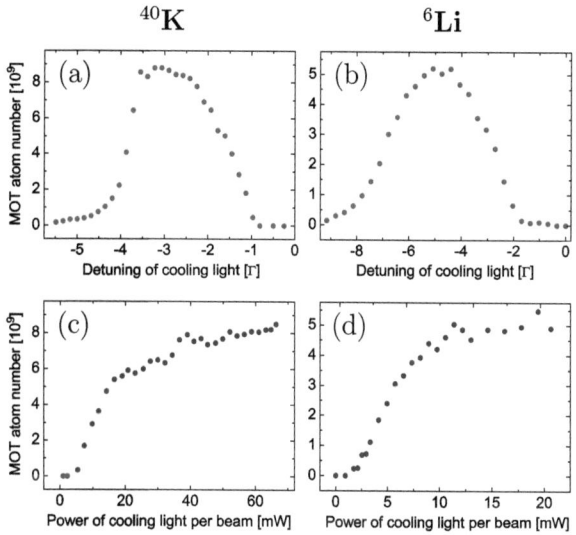

Figure 3.8: MOT atom number as a function of (a,b) the detuning and (c,d) the power of the cooling light per MOT beam for a constant intensity ratio between the cooling and repumping light. The graphs (a,c) relate to the ^{40}K-MOT and (b,d) to the ^{6}Li-MOT. For ^{40}K a power of $45\,$mW corresponds to an intensity of $13\,I_\text{sat}$, for ^{6}Li a power of $20\,$mW corresponds to an intensity of $4\,I_\text{sat}$, with the respective saturation intensities I_sat given in Tab. 2.3.

Figures 3.8 (c,d) show the dependence of the MOT atom number on the power of the cooling light per MOT beam. In the figures, a power of $10\,$mW corresponds to an on-resonance peak intensity of $\sim 3\,I_\text{sat}$ (Fig. 3.8 (c)) and $\sim 2\,I_\text{sat}$ (Fig. 3.8 (d)) in each of the six MOT beams. The atom number increases with increasing light power and saturates for higher powers. The saturation is due to several effects. First, the optical transition used for cooling saturates for high intensities. Second, the repulsive forces between the atoms due to rescattered photons and the temperature of the cloud increase with light power [163]. Finally the scattering rate for light-induced cold collisions increases with light power.

The dependence of the MOT atom number on the axial magnetic field gradient is shown in Fig. 3.9. For the ^{40}K-MOT the atom number has a maximum for a gradient of $8\,\text{G/cm}$. This gradient is thus chosen for the dual-species operation. The figure shows that the atom number in the ^{6}Li-MOT is not optimum for this value, it is larger for higher gradients.

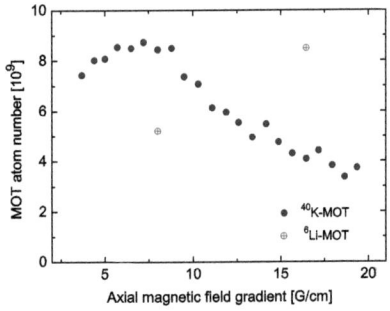

Figure 3.9: MOT atom number as a function of the axial magnetic field gradient. The atom number in the ^{40}K-MOT is maximum for a gradient of $8\,\text{G/cm}$. This gradient is chosen for the dual-species operation. The ^{6}Li-MOT contains more atoms for higher magnetic fields.

Figure 3.10 (a) shows the dependence of the ^{40}K-MOT atom number on the intensity ratio $I_{\text{rep}}/I_{\text{cool}}$ between repumping and cooling light for three different repumping detunings $\Delta\omega_{\text{rep}}^{(1)} = -3\,\Gamma$, $\Delta\omega_{\text{rep}}^{(2)} = -5\,\Gamma$ and $\Delta\omega_{\text{rep}}^{(3)} = -7\,\Gamma$ and a constant cooling light power of $18\,\text{mW}$ per MOT beam. The curves have a maximum at different ratios $I_{\text{rep}}/I_{\text{cool}}$, the position of the maxima lying at higher ratios for lower detunings. Furthermore, the maxima have different values for the three curves. The maximum is biggest for the detuning $\Delta\omega_{\text{rep}}^{(2)} = -5\,\Gamma$. The shape of the curves can be understood as follows. Each curve increases between $I_{\text{rep}}/I_{\text{cool}} = 0$ and the position of the maximum, because the transition probability of the repumping transition increases with increasing repumping intensity. Thus the atoms are more efficiently cooled by the cooling light, as they are more efficiently repumped into the cycling transition. However, when the intensity of the repumping light becomes too large, the curve decreases again. Then, due to the strong repumping, the atoms are exposed to the more intense near-resonant

3.3. ^{6}Li-^{40}K dual-species MOT

cooling light, which causes light-induced cold collisions, leading to trap loss. At the maximum, the repumping is sufficiently strong to allow for an efficient cooling, and it is sufficiently weak to preserve the atoms from cold collisions induced by the strong cooling light. The value of the curve maximum is biggest for the detuning $\Delta\omega_{\text{rep}}^{(2)} = -5\,\Gamma$. It is situated at $I_{\text{rep}}/I_{\text{cool}} \sim 1/20$, for which, as one can see below, only $\sim 20\%$ of the ^{40}K-MOT atoms occupy the cooling cycle states $F = 9/2$ or $F' = 11/2$ (see Fig. 3.10 (b)), the others occupying the "dark" hyperfine ground state $F = 7/2$.

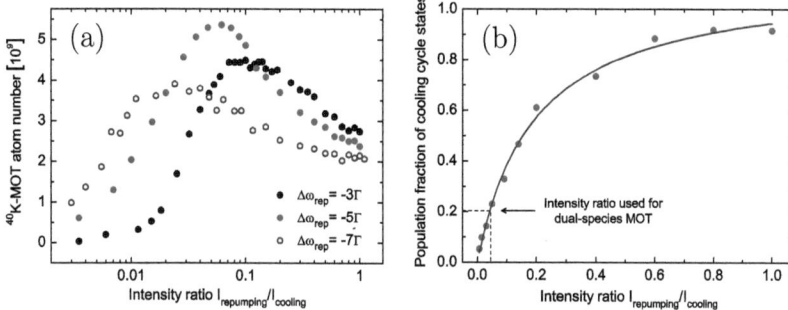

Figure 3.10: (a) ^{40}K-MOT atom number as a function of the intensity ratio between repumping and cooling light for three different repumping detunings $\Delta\omega_{\text{rep}}$ and a constant cooling light power of 18 mW per MOT beam (which corresponds to an intensity of $6\,I_{\text{sat}}$). (b) Circles: measured fraction of atoms in the ^{40}K-MOT populating the states $F = 9/2$ or $F' = 11/2$ (cooling cycle states) as a function of the intensity ratio between repumping and cooling light for the repumping detuning $\Delta\omega_{\text{rep}} = -5\,\Gamma$ and a constant cooling light power of 18 mW per MOT beam. For the ratio which maximizes the total atom number in the ^{40}K-MOT, $I_{\text{rep}}/I_{\text{cool}} \sim 1/20$, only 20% of the trapped atoms occupy the cooling cycle states. Solid curve: a fit based on Einstein's rate equations.

For very small intensity ratios $I_{\text{rep}}/I_{\text{cool}} \leq 0.01$ the atom number in the ^{40}K-MOT is larger for higher repumping detunings (Fig. 3.10 (a)). This behavior might be a consequence of the fact that the ^{40}K-MOT is loaded from a slow atomic beam. The beam atoms, which have a negative Doppler shift of more than $5\,\Gamma$ with respect to the counter-propagating MOT beams, might absorb the repumping light more likely when it has a higher detuning.

Figure 3.10 (b) shows the fraction of atoms in the ^{40}K-MOT (circles) which populate the states $F = 9/2$ or $F' = 11/2$ (i.e. the cooling cycle states, see Fig. 2.3) as a function of the intensity ratio $I_{\text{rep}}/I_{\text{cool}}$ between repumping and cooling light. In the experiment, the cooling light power was fixed to 18 mW per MOT beam, and the repumping detuning was $\Delta\omega_{\text{rep}} = -5\,\Gamma$. The graph was recorded as follows. The absolute population of the states $F = 9/2$ and $F' = 11/2$ was measured by simultaneously switching off both the repumping and cooling light of the ^{40}K-MOT 600 μs before taking the image (with the imaging beam being near-resonant with the $F = 9/2 \to F' = 11/2$-transition). During the 600 μs time delay, all excited atoms relax to one of the ground states. For the used detunings and intensities of the MOT-beams $\sim 90\%$ of the excited atoms occupy the state $F' = 11/2$ and thus relax to the ground state $F = 9/2$, which is imaged. Therefore, the image approximately yields the sum of the populations of the states $F = 9/2$ and $F' = 11/2$. The total population of all states (*i.e.* the total number of trapped atoms) was measured as usual by first optically pumping all atoms into the hyperfine ground state $F = 9/2$ before taking the image.

The curve in Fig. 3.10 (b) is increasing with increasing ratios $I_{\text{rep}}/I_{\text{cool}}$ and it saturates for high ratios. For the ratio $I_{\text{rep}}/I_{\text{cool}} = 1/5$ about 60% of the ^{40}K-MOT atoms occupy the cooling cycle states. For this ratio the fluorescence emitted by the ^{40}K-MOT is found to be maximum. For the ratio $I_{\text{rep}}/I_{\text{cool}} = 1/20$, which is used in the experiment, only $\sim 20\%$ of the atoms occupy the cooling cycle states. Atom losses due to light-induced collisions are thus reduced.

The solid curve in Fig. 3.10 (b) shows a fit of the experimental data, based on a simple model, assuming ^{40}K to be a four-level atom (with the states $F = 9/2$, $F = 7/2$, $F' = 11/2$ and $F' = 9/2$). Einstein's rate equations yield that the curve obeys the law $P_{\text{ccs}} = 1/(1 + a + b/(I_{\text{rep}}/I_{\text{cool}}))$, with the fitting parameters $a = -0.1$ and $b = 0.17$, which depend on the transition probabilities and the used intensities and detunings. Even though the model describes well the measured data, the model is limited, since $P_{\text{ccs}} \to 1.1$ for $I_{\text{rep}}/I_{\text{cool}} \to \infty$ and thus serves here merely to guide the eye.

Figure 3.11 shows the dependence of the ^{6}Li-MOT atom number on the intensity ratio $I_{\text{rep}}/I_{\text{cool}}$ between repumping and cooling light for the repumping detuning $\Delta\omega_{\text{rep}} = -3\,\Gamma$ and a constant cooling light power of 11 mW per MOT beam. In contrast to Figure 3.10 (a), the curve does not have a maximum but rather increases with increasing $I_{\text{rep}}/I_{\text{cool}}$ and saturates. This behavior is a result of the important contribution of the repumping light to the cooling process, particular to ^{6}Li, as it has an unresolved excited-state hyperfine structure.

3.3.2 Heteronuclear Collisions in the dual-species MOT

We now turn to the study of heteronuclear collisions in the dual-species MOT. We quantified the homo- and heteronuclear collision rates and studied the dependence of heteronuclear collision rates on the laser power used for the MOT-beams.

In a dual-species MOT, inelastic collisions between atoms of the two different species

3.3. ^{6}Li-^{40}K dual-species MOT

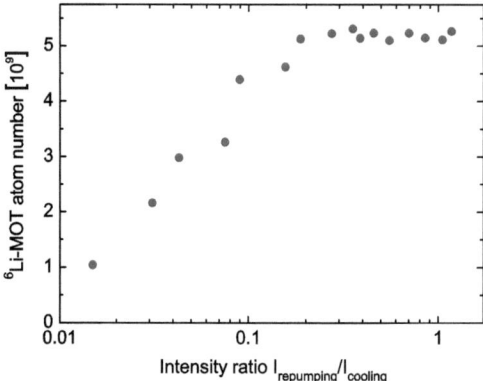

Figure 3.11: ^{6}Li-MOT atom number as a function of the intensity ratio between repumping and cooling ligh for a constant cooling light power of 11 mW per MOT beam (which corresponds to an intensity of $2\,I_{\text{sat}}$). In comparison to ^{40}K (Fig. 3.10 (a)), the optimum atom number requires a larger intensity in the repumping light, which is a consequence of the unresolved excited hyperfine structure of ^{6}Li.

can occur and represent an important loss mechanism. Previous studies have shown that the principal loss mechanisms for heteronuclear collisions in dual-species MOTs involve one ground-state and one excited atom of different species [146, 147]. Such atom pairs can undergo radiative escape or fine-structure changing collisions [165]. Both these loss processes require the two atoms to approach each other sufficiently close such that a large enough interaction energy is gained to make the atoms leave the trap. The long-range behavior of the scattering potentials determines if the atoms can approach each other sufficiently. For LiK, the scattering potentials for a singly excited heteronuclear atom pair are all attractive for the case where the K atom is excited, on the contrary they are all repulsive for the case where the Li atom is excited [166]. As a consequence, a ground-state K atom and an excited Li atom repel each other and are prevented from undergoing inelastic collisions (optical shielding). Inelastic collisions involving singly excited heteronuclear atom pairs thus always contain an excited K atom. In order to minimize the rate of heteronuclear collisions in the LiK-MOT, the density of excited K atoms must therefore be reduced. Furthermore, the atomic density in the trap as well as the relative speed of the colliding atoms, *i.e.* the temperature of the cloud, need to be reduced.

In our ^{6}Li-^{40}K dual-species MOT the following strategy is applied in order to minimize inelastic heteronuclear collisions. First the use of low magnetic field gradients (8 G/cm), which decreases the atomic densities ($n_c^K \sim 3 \times 10^{10}$ atoms/cm^3 and

$n_c^{\text{Li}} \sim 2 \times 10^{10}$ atoms/cm³). Second, low intensities in the repumping light for both, ⁶Li and ⁴⁰K, are used in order to decrease the number of excited atoms. Decreasing the number of excited ⁶Li atoms here *a priori* serves to decrease the temperature of the ⁶Li-cloud. Since that is much larger than the temperature of the ⁴⁰K-cloud, the relative speed of two colliding atoms and thus the collision rate can be efficiently decreased by minimizing the temperature of the ⁶Li-cloud. Finally a small mutual influence of the MOTs is obtained: the atom numbers in the MOTs decrease by $\sim 4\%$ in the ⁶Li-MOT and $\sim 10\%$ in the ⁴⁰K-MOT due to the presence of the other species.

The importance of decreasing the magnetic field gradients in order to minimize the heteronuclear collision rate in the dual-species MOT is demonstrated in Fig. 3.12 (a), which depicts the effect of the ⁶Li-MOT on the ⁴⁰K-MOT atom number when a two-times larger magnetic field gradient (16 G/cm) is used. At this gradient the atomic density in the ⁶Li-MOT is by a factor of 4 larger than at the gradient used for the optimized MOT. In the experiment, the ⁴⁰K-MOT was intentionally reduced in size (by decreasing the 2D-MOT flux) to ensure a better inclosure in the ⁶Li-MOT. The curve shows that $\sim 65\%$ of the ⁴⁰K-MOT atoms leave the trap due to the enhanced heteronuclear collisions. Using a low magnetic field gradient is therefore helping significantly to decrease the heteronuclear collisions.

In the following we determine the trap loss coefficients for the (optimized) dual-species MOT in order to quantify the heteronuclear collisions. The rate equation for the atom number in a dual-species MOT (with species A and B) reads [146]

$$\frac{dN_A}{dt} = L_A - \gamma N_A - \beta_{AA} \int n_A^2 dV - \beta_{AB} \int n_A n_B dV, \qquad (3.12)$$

where L_A is the loading rate, γ the trap loss rate due to collisions with background gas atoms and n_A, n_B the local atomic densities. β_{AA} and β_{AB} denote the cold collision trap loss coefficients for homo- and heteronuclear collisions, respectively. L_A and γ are determined from the loading and decay curves of the single-species MOTs. The obtained values for L_A are given in Tab. 2.3 and γ is found to be $0.13\,\text{s}^{-1}$. The homonuclear trap loss coefficients β_{AA} are determined from the steady state atom numbers in single-species operation using the measured density profiles. For the experimental conditions indicated in Tab. (2.3), we obtain

$$\beta_{\text{LiLi}} = (8 \pm 4) \times 10^{-12}\,\text{cm}^3\text{s}^{-1}, \qquad (3.13)$$
$$\beta_{\text{KK}} = (6 \pm 3) \times 10^{-13}\,\text{cm}^3\text{s}^{-1}. \qquad (3.14)$$

The determination of the heteronuclear trap loss coefficients β_{AB} for the optimized

3.3. ^{6}Li-^{40}K dual-species MOT

dual-species configuration would require the knowledge of the mutual overlap of the MOTs, which is difficult to estimate when absorption images are taken only along one direction. We therefore choose a configuration, which makes the determination of β_{AB} less dependent on assumptions about the mutual overlap (but which does not change the value of β_{AB}). We reduce the atom flux of species A, in order to decrease the spatial extension of the trapped cloud of species A and to place it in the center of the cloud of species B. A video camera which records the fluorescence of the MOTs from a different direction than that of the absorption imaging verifies that this configuration is indeed achieved. Then, in Eq. (4.21) it is $\int n_A n_B dV \sim n_c^B N_A$. Comparing the steady-state atom numbers for the different configurations then yields

$$\beta_{\text{LiK}} = (1 \pm 0.5) \times 10^{-12}\,\text{cm}^3\text{s}^{-1}, \quad (3.15)$$

$$\beta_{\text{KLi}} = (3 \pm 1.5) \times 10^{-12}\,\text{cm}^3\text{s}^{-1}, \quad (3.16)$$

for the experimental conditions indicated in Tab. (2.3). Comparing all four trap loss coefficients, the dominant is β_{LiLi} (Eq. (3.13)) for light-induced homonuclear ^{6}Li-^{6}Li collisions. This is a consequence of the large temperature of the ^{6}Li-MOT and the unresolved hyperfine structure of ^{6}Li which prohibits the creation of a dark MOT, leading to a large excited-state population. The much smaller homonuclear trap loss coefficient β_{KK} for ^{40}K (Eq. (3.14)) is consistent with Fig. 3.10 (a) which shows that, for ^{40}K, small repumping intensities are favorable. The heteronuclear trap loss coefficients $\beta_{\text{LiK}}, \beta_{\text{KLi}}$ (Eqs. (3.15) and (3.16)) are also much smaller than β_{LiLi}, indicating that our applied strategy for decreasing the heteronuclear collisions is good. In the Amsterdam group the heteronuclear trap loss coeffiecients were found by a factor of about 2 larger than ours [44]. A dark SPOT MOT has been implemented in order to reduce the excited-state population of the ^{40}K atoms. In the next paragraph we show, however, that it is also important to reduce the excited-state population of the ^{6}Li atoms.

Figure 3.12 (b) depicts the dependence of the trap loss coefficient β_{KLi} on the relative excited-state population of the ^{6}Li atoms. The graph was obtained by recording the influence of the ^{6}Li-MOT on the ^{40}K-MOT as the power of the ^{6}Li-MOT beams was varied. For each power it was verified that the ^{40}K-MOT was placed in the center of the ^{6}Li-MOT and the atomic density of the ^{6}Li-MOT was recorded. In the experiment a magnetic field gradient of 16 G/cm was used. The central atomic density of the ^{6}Li-MOT was found to be approximately constant, when the power was varied ($n_c^{\text{Li}} \sim 8 \times 10^{10}$ atoms/cm^3). The relative excited-state population for a given beam power was estimated using Einstein's rate equations (assuming ^{6}Li to be a three-level atom with the states $2S(F=1/2)$, $2S(F=3/2)$ and $4P_{3/2}$). The predicted variation

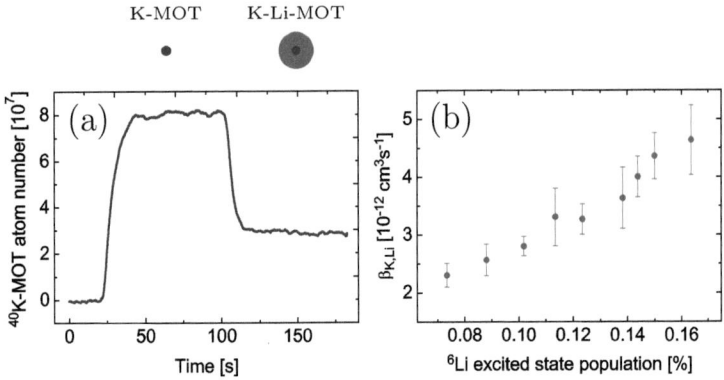

Figure 3.12: (a) Evolution of the atom number in the ^{40}K-MOT in the absence ($t < 100\,\text{s}$) and presence ($t > 100\,\text{s}$) of the ^{6}Li-MOT for an increased magnetic field gradient of $16\,\text{G/cm}$. (b) Trap loss coefficient β_{KLi} for heteronuclear collisions as a function of the relative excited-state population of the trapped ^{6}Li atoms.

of the excited-state population with the beam power was in accordance with the measurements of the fluorescence emitted by the ^{6}Li-MOT and the number of captured atoms via absorption imaging. The latter changed by a factor of 1.5 in the considered range of beam powers. The graph in Fig. 3.12 (b) shows that the trap loss coefficient increases by more than a factor of 2 as the relative excited-state population is increased from $\sim 7\%$ to $\sim 16\%$. The error bars shown in the figure refer to statistical errors. The uncertainty due to systematic errors is estimated to be 50%. The significant increase of β_{KLi} demonstrates the importance of minimizing the number of excited ^{6}Li atoms (and not only that of the excited ^{40}K atoms). One reason for this increase is the increase of temperature of the ^{6}Li-MOT, which changes from $\sim 1\,\text{mK}$ to $\sim 1.6\,\text{mK}$ when the beam power is increased. Another reason could be the occurence of collisions involving doubly excited Li*K* atom pairs, the rate of which increases with the excited-state populations. The scattering potentials for these collisions are known to be of a long-range, as they scale with the internuclear separation as $1/R^5$ [167], whereas they scale as $1/R^6$ for collisions involving a singly excited heteronuclear atom pair [164]).

3.4 Transfer of the atoms into the magnetic trap

We now study the transfer of the atoms from the MOT to the magnetic quadrupole trap of the MOT chamber. After a compressed MOT stage, the atoms are optically pumped into their stretched states by a short pulse of resonant circularly polarized

3.4. Transfer of the atoms into the magnetic trap

light. In the following we characterize this optical pumping stage in more detail. We focus here on the case of ^{40}K.

Figure 3.13 depicts the number of ^{40}K atoms in the magnetic quadrupole trap as a function of time for the cases where optical pumping is applied before the transfer and where not. In the latter case, the number of atoms drastically decreases within the first ~ 20 ms by a factor of ~ 10. The atoms which escape from the trap during this time are either in a spin state which is repelled from the trap or which is susceptible to undergo spin relaxation. The fast atom number reduction stops when only atoms in stable spin states are left. Then, the atom number continues to decrease, but on a much larger time scale, which is due to collisions with background gas atoms. In the case where optical pumping is applied, nearly no atoms are lost within the first 20 ms and the decay of the atom number only happens on a large time scale identical to the one of the previous case. The figure demonstrates the importance of the optical pumping stage for the atom transfer efficiency. Furthermore it shows that the transfer efficiency in our experiment for the case of ^{40}K is close to 100%.

We expect that the optimized optical pumping stage prepares nearly all ^{40}K atoms in their stretched state. However, since we intend to evaporatively cool ^{40}K in the magnetic trap of the final cell, the atoms will have to be prepared in at least two spin states with significant populations, requiring to optically pump less efficiently.

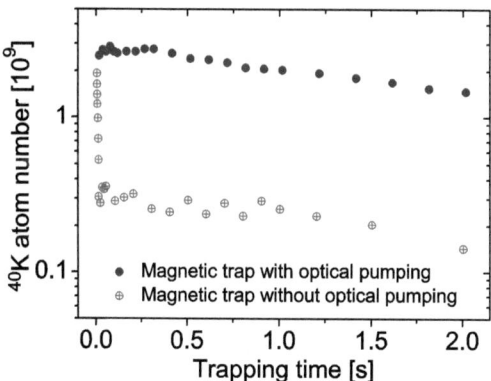

Figure 3.13: Decay curves of the ^{40}K magnetic trap for the case where optical pumping is applied before the atoms are transferred from the MOT to the magnetic trap and for the case where no optical pumping is applied.

In order to optimize the optical pumping stage for ^{40}K, we measured the number of atoms transfered to the magnetic trap (measured after one second of trapping) for various parameters, such as: the pulse duration, the intensity of the two frequency components (principal and repumper), the value of the applied bias magnetic field, the quality of the beam's circular polarization and the quality of the power balance between the two counter propagating optical pumping beams. We found that the optical pumping stage heats up the atom cloud. We were able to minimize this heating by shortening the pulse duration to $5\,\mu$s (short compared to the $200\,\mu$s-duration used for ^{6}Li). This short duration requires a large beam intensity. Saturation of the transfer efficiency was measured to occur for intensities above $\sim 30\, I_{\text{sat}}$ per beam. The intensity of the repumping light, which is responsible for pumping the atoms in the correct hyperfine state, was found to saturate above $\sim 5\, I_{\text{sat}}$ per beam. The large light power of the optical pumping beams is obtained by deriving them from the light beam used for the ^{40}K-MOT: the MOT beam passes through a single pass AOM, the first order of which is used for the MOT. During the optical pumping phase, this AOM is switched off, providing power in the zeroth order, which is used for the optical pumping. For ^{6}Li we use a similar strategy: The MOT beam passes through a single pass AOM, the zero order of which is used for the MOT. During the optical pumping phase, this AOM is switched on, providing power in the first order, which is used for the optical pumping.

Another important parameter for the optical pumping is the value of the bias magnetic field. Its optimum value was found to be ~ 2 G. Larger values lead to a decrease in the optical pumping efficiency, since then the Zeeman effect detunes the different repumping transitions. Even in the absence of the bias magnetic field the optical pumping is quite efficient: the transfer efficiency then is only $\sim 50\%$ less than its optimum value. This is probably due to the large intensity of the optical pumping beam, which can overcome the depolarizing effect of small stray magnetic fields. It was found to be important to power balance the optical pumping beam by shining it with equal intensities from two opposite directions on the atoms. Not only an increase of $\sim 30\%$ in transfer efficiency could thus be gained, but also a significant decrease of the heating which is created by the light pulse. We still observe a small remaining heating effect, which is minimized by reducing the used light power.

3.5 Magnetic quadrupole trap

We now investigate the trapping of atoms in the magnetic quadrupole trap. Figure 3.14 shows the time evolution of the number of atoms in both the magnetic trap and the MOT for the two different atomic species. For each data point the respective trap was prepared under the same conditions and the atom number was determined by absorption images which were taken for different times after the end of the trap loading. The decay of the number of atoms due to collisions with background gas atoms in either

3.5. Magnetic quadrupole trap

of the trap is approximately described by the equation

$$\frac{dN}{dt} = -\gamma N, \qquad (3.17)$$

with the lifetime of the atoms defined as $\tau = 1/\gamma$. Fitting the data in Fig 3.14 to Eq. (3.17) yields the lifetimes of the different atoms and traps. Since in a MOT additional light-induced cold collisions can take place, the decay curves for the MOTs are fitted for larger trapping times only, for which these collisions are insignificant (this was only done for the ^{6}Li-MOT, since for the ^{40}K-MOT light-induced collisions are negligible due to the small trap loss coefficient β_{KK} (see Eq. (3.14))). The fits yield the lifetimes $\tau^K_{MT} = 3.1\,\text{s}$, $\tau^K_{MOT} = 7.5\,\text{s}$ and $\tau^{Li}_{MT} = 3.0\,\text{s}$, $\tau^{Li}_{MOT} = 7.4\,\text{s}$, respectively. Thus, about the same lifetimes are obtained for the two species, but significantly different ones for the two different traps. Before investigating this difference we first discuss the total value of the measured lifetimes.

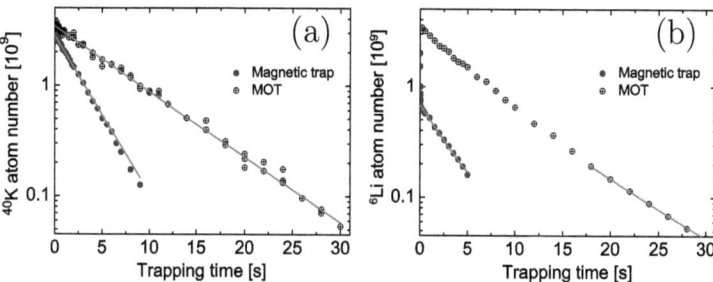

Figure 3.14: Decay curves of an atom cloud trapped in the magnetic quadrupole trap in comparison to the decay curves obtained for trapping in the MOT for (a) ^{40}K and (b) ^{6}Li. Dots represent the experimental data and solid lines represent fits to Eq. (3.17) from which the respective trap lifetimes $\tau^K_{MT} = 3.1\,\text{s}$, $\tau^K_{MOT} = 7.5\,\text{s}$ and $\tau^{Li}_{MT} = 3.0\,\text{s}$, $\tau^{Li}_{MOT} = 7.4\,\text{s}$ are obtained. The atom number in the magnetic trap for ^{6}Li decreases rapidly within the first 200 ms due to an imperfect optical pumping stage.

The measured lifetimes are of the order of 5 s, which is relatively low, given the fact that the MOT chamber is isolated from the atom sources. In a previous setup, which we had used before enlarging the vacuum system by the magnetic transport section, we could obtain lifetimes of the order of 30 s in the MOT chamber using the same atom sources. The smaller lifetime of the new setup is not due to a leak, which we verified by a helium leak test, neither is it due to an insufficient isolation of the atom sources, as was verified making use of the valves connecting them to the MOT chamber. We believe that it is due to a contamination of the added homemade parts which have been soldered to the octagonal MOT cell. Probably the cleaning has not been done sufficiently well after the soldering at the machine shop. The low atomic lifetimes in the MOT chamber unfortunately affect the lifetime in the final cell. We have measured the atomic lifetime in the magnetic quadrupole trap in the final cell, which was found to be $\sim 26\,\text{s}$. We hope that this value is large enough to allow for a sufficiently large ratio between elastic to inelastic collisions necessary for evaporative cooling of the two atomic species.

We now address the question why the lifetime in the two traps is significantly different. The measurements yield that the lifetime in the MOTs is larger by a factor of ~ 2 than in the magnetic traps. It can be excluded that the smaller lifetime in the magnetic trap is due to spin relaxation collisions between the trapped atoms, since those happen during the first 200 ms, leaving behind only atoms which are not susceptible to those collisions. Also Majorana losses can be excluded as the origin of the smaller lifetime due to the large temperature of the atoms. We can further exclude that the larger lifetime in the MOTs results from a non-negligible loading from the background vapor, since we do not load a detectable number of atoms when the atom sources are switched off. Differences between atomic lifetimes in MOTs and magnetic traps have been reported and studied by several groups [163, 168, 169]. It has been argued that the different lifetimes are the result of the different depths of the traps, which lead to a different probability for recapture of the collisional products. In our setup, the trap depth of the MOTs is $\sim 1\,\text{K}$ and the trap depth of the magnetic trap is $\sim 15\,\text{mK}$ and thus about two orders of magnitude less. Small-angle elastic collisions between the background and the trapped atoms are thus more likely to lead to trap loss for the magnetic trap than for the MOT.

In order to allow for a quantitative comparison between the theoretically predicted lifetimes which result from the different trap depths, we estimate the cross section for knock-out collisions between a room-temperature background gas atom and a trapped cold atom. The estimation is based on the approach described in Ref. [163]. The colliding partners interact with each other via the van der Waals interaction which has a scattering potential of the form $U(R) = -C_6^{\text{eff}}/R^6$, where R is the interatomic distance and C_6^{eff} an effective dispersion coefficient accounting for all the different atomic species present in the background vapor. The trapped atom will be knocked out of the trap if its velocity gain due to the collision exceeds the escape velocity v_{esc} of the trap. We now calculate the impact parameter b, for which the velocity gain just equals v_{esc}, the cross section for knock-out collisions is then given by $\sigma = \pi b^2$. Since the velocity of the trapped atoms is small compared to the velocity of the background atoms, the background atom is likely to be hardly affected by the collision and can be considered as moving at constant velocity along a straight line as illustrated in Fig. 3.15. The change in velocity of the trapped atom then is

$$\Delta \boldsymbol{v} = \int_{-\infty}^{\infty} \boldsymbol{F}/m \, \mathrm{d}t, \qquad (3.18)$$

where

$$F = \frac{6C_6^{\text{eff}}}{R^7}, \quad R = \frac{b}{\cos\theta}, \quad \tan\theta = \frac{vt}{b} \Leftrightarrow \mathrm{d}t = \frac{b}{v}\frac{1}{\cos^2\theta}\,\mathrm{d}\theta, \qquad (3.19)$$

where v denotes the speed of the background atom. The component of $\boldsymbol{F}$ along the flight direction of the fast atom averages to zero and the integral over the remaining component, $F\cos\theta$, is given by

$$\Delta v = \int_{-\pi/2}^{\pi/2} \frac{6C_6^{\text{eff}}}{mvb^6}\cos^6\theta\,\mathrm{d}\theta = \frac{15\pi}{8}\frac{C_6^{\text{eff}}}{mvb^6}, \qquad (3.20)$$

Thus, the relations $\Delta v = v_{\text{esc}}$ and $\sigma = \pi b^2$ yield

$$\sigma = \pi \left(\frac{15\pi C_6^{\text{eff}}}{3mv_{\text{esc}}v}\right)^{1/3}. \qquad (3.21)$$

The scattering cross section is, according to kinetic gas theory, related to the lifetime via $\tau = 1/(\sigma n_b \bar{v})$, where n_b is the density of background atoms and $\bar{v}$ their mean velocity (considering the velocity of the trapped atoms to be zero). In our system the escape velocities of the MOTs are $v_{\text{esc}}^{\text{KMOT}} \sim 15\,\text{m/s}$ and $v_{\text{esc}}^{\text{LiMOT}} \sim 30\,\text{m/s}$ and the escape velocities of the magnetic traps are $v_{\text{esc}}^{\text{KMT}} \sim 1.7\,\text{m/s}$ and $v_{\text{esc}}^{\text{LiMT}} \sim 4.2\,\text{m/s}$, respectively. According to Eq. (3.21), we thus expect a difference in lifetimes for the two traps of

$$\frac{\tau_{\text{MOT}}}{\tau_{\text{MT}}} = \left(\frac{v_{\text{esc}}^{\text{MOT}}}{v_{\text{esc}}^{\text{MT}}}\right)^{1/3} \sim 2, \qquad (3.22)$$

which is approximately the same for both species and in accordance with the observed value of ~ 2.

3.6 Magnetic transport

The next step in the experimental sequence which follows the magnetic trapping is a magnetic transport to the science chamber. We have achieved the transport of both atomic species with transport efficiencies of $\sim 15\%$ for ^{40}K and $\sim 7\%$ for ^{6}Li. We

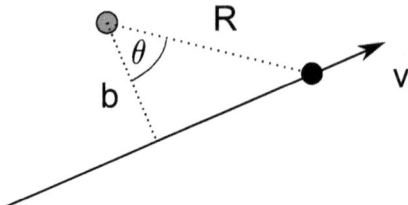

Figure 3.15: Illustration of a collision between a room-temperature background gas atom and a trapped atom. Due to the much larger velocity of the background gas atom, that is nearly not deviated due to the collision.

believe that we will be able to achieve higher transport efficiencies in future. We found that the transport efficiency of our system is limited by the small size of the vacuum tubes through which the transport path is directed. Those have a small diameter of 1 cm and thus require a very good centering of the transport path with respect to the tube's axis. Furthermore, the atom clouds need to have small sizes, which requires small cloud temperatures and a strong confinement of the atomic clouds. We have recently purchased new current supplies which are able to deliver higher currents in order to increase the confinement. In the following we briefly describe the cruicial steps which we followed in order to make the magnetic transport operational.

Before the transport sequence was run for the first time, the setup needed to be debugged. The first step is to verify the agreement between the demanded and the observed current flows through the coils. In particular it has to be verified that no significant time delays and possible fluctuations of those exist between the two signals. For the typical transport durations of ~ 3 s and used accelerations of $a \sim 0.75\,\mathrm{m/s^2}$, maximum velocities of the potential minimum position of $\sim 50\,\mathrm{cm/s}$ are obtained. So not to induce a displacement of more than 1 mm between the actual and the demanded position, the time delay between the actual and demanded coil currents should be less than 2 ms. These small time delays require the current supplies to be operated in constant voltage mode. Another important initial check consisted in verifying that no spikes in the currents were created due to the switching between the coils or the change of operation mode of the power supplies. The calculated current waveforms sometimes also needed to be checked, as numerical artifacts in the calculation might introduce spikes in the waveform, which then need to be manually corrected in the waveform file. In order to debug the switching mechanism for the coil currents, we installed a series of LEDs which indicate when current is flowing in the different coil pairs. Each employed LED has a threshold voltage of 1.6 V and is connected in parallel to one of the coil pairs together with a 1 kΩ resistor. The voltage drop between the coil ends, which is present when current flows through the coils, then switches on the LED. The switching mechanism for the coil currents is realized with a series of MOSFETs, which are fragile components that occasionally die. The LED system was also found helpful to immediately identify the MOSFETs which need to be replaced.

The initial optimization of the transport was done by transporting the atom cloud forth and back over varying distances. This configuration allows the precise determination of the transport efficiencies for each transport distance, since the same imaging system for the atom number determination before and after the transport can be used. Furthermore it allows us to probe the critical positions on the transport path (such as the positions where the aspect ratio of the trapping potential changes or where the vacuum tubes decrease their diameter).

An important parameter of the transport is its total duration. The optimum duration is a compromise between heating of the atom cloud due to nonadiabatic acceleration or changes of the aspect ratio of the potential, and atom loss due to collisions of the transported atoms with the background gas. Since the background gas pressure changes along the transport path, the efficiency of the transport can be optimized by adjusting the transport velocity to the local pressures. The transport path consists of two parts, from the MOT to the elbow and from the elbow to the final cell. Since the lifetime in the MOT chamber is much less ($\sim 3\,\text{s}$) than in the science cell ($\sim 26\,\text{s}$), it is preferable to transport the cloud fast in the first part of the transport and less fast in the second part. In the experiment we have used an acceleration profile for which the acceleration magnitude is constant for each of the two parts of the transport and twice as large in the first part than in the second. The optimum transport time for this profile was then found to be 2.7 s for both atomic species. Figure 3.16 shows the dependence of the number of transported ^{40}K atoms as a function of the transport time for the described acceleration profile.

The optimization of the magnetic transport is still in progress. We are convinced that with a stronger magnetic confinement, which will be made possible by using the new current supplies, and with a fine-tuning of the alignment of the transport path we will be able to significantly improve the transport efficiency. Once the atom number in the final cell is sufficiently large we will start evaporating the atomic clouds in the plugged magnetic trap.

3.7 Conclusion

In this chapter we have presented a characterization of the implemented experimental apparatus. We have demonstrated a very satisfying performance of the constructed

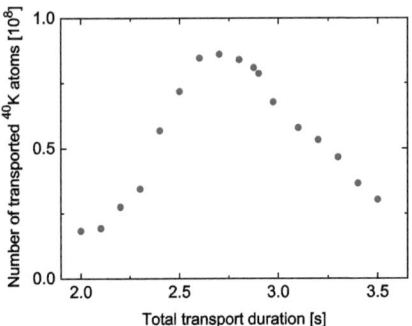

Figure 3.16: Number of magnetically transported atoms as a function of the total transport duration. The initial atom number in the magnetic trap before the transport was $\sim 5 \times 10^8$ at the day when the graph was recorded.

atom sources, and the dual-species magneto-optical trap. The performance of the magnetic transport still requires some more optimization, but we are convinced that the constructed apparatus will allow the production of quantum degenerate Fermi-Fermi mixtures with very large atom numbers in close future.

In the following we summarize the main characteristics of the different subsystems.

- The **^{6}Li Zeeman slower** yields a capture rate of $\sim 1.2 \times 10^9$ atoms/s in the ^{6}Li-MOT at an oven temperature of 500 °C. A higher oven temperature would yield a higher capture rate, however, we limit it to 500 °C, since, as we have shown, collisions between the atoms in the atomic beam with the atoms in the MOT become important for higher temperatures. The atomic beam is diverging, having a radius of ~ 5 cm at the position of the MOT. The bichromatic laser beam of the slower has a total power of 50 mW. It is converging toward the oven, having a $1/e^2$-diameter of 3 cm at the position of the MOT and of 0.8 cm at the position of the oven's collimation tube. The used light power saturates the ^{6}Li-MOT capture rate. 1/10-th of the light power is used for repumping.

- The **^{40}K 2D-MOT** yields a capture rate of $\sim 1.4 \times 10^9$ atoms/s in the ^{40}K-MOT. The mean velocity of the atomic beam was measured to be ~ 24 m/s. A significant amount of the available light power was found to be required for repumping (1/3-rd of the power). The capture rate was found optimum for a potassium vapor pressure of 2.3×10^{-7} mbar in the 2D-MOT cell. For higher pressures collisions between the vapor atoms and the atoms in the atomic beam become important, decreasing the capture rate. In contrast to 2D-MOT systems of other species, we found that the limiting collisions in our system are collisions

3.7. Conclusion

between non-excited ^{40}K and non-excited ^{39}K atoms. An analysis of our data showed that the capture rate could thus be improved by up to a factor of 10 by using a potassium sample of a higher isotopic enrichment.

- The **^{6}Li-^{40}K dual-species MOT** simultaneously traps 5.2×10^9 ^{6}Li atoms and 8.0×10^9 ^{40}K atoms. The use of small magnetic field gradients and low light powers in the repumping light for both species were found important to minimize intra- and interspecies light-induced collisions. The low light powers in the repumping light lead to a small fraction of excited atoms. We have measured the different trap loss coefficients and found that for our experimental parameters the trap loss coefficient for intraspecies ^{6}Li collisions is dominant and about 4 times larger than the trap loss coefficients for interspecies and intraspecies ^{40}K collisions. The large trap loss coefficient for intraspecies ^{6}Li collisions results from the need to use a relatively large repumping power for the ^{6}Li-MOT due to the small excited-state hyperfine structure. We have studied the dependence of one of the interspecies trap loss coefficients on the number of excited ^{6}Li* atoms. Since the scattering potential for singly excited ^{6}Li*-^{40}K collisions is repulsive at long-range, these collisions are not expected to lead to trap loss. Nonetheless we found that the interspecies trap loss coefficient depends on the excited-state population of the ^{6}Li atoms, indicating that maybe collisions involving doubly excited Li*-K* atom pairs occur.

- The **magnetic trap** in the MOT chamber is loaded efficiently with the atoms from the MOT by applying a compressed MOT and an optical pumping stage. Nearly 100% of the ^{40}K atoms and $\sim 30\%$ of the ^{6}Li can be transferred into the magnetic trap. For ^{6}Li the efficiency is smaller, since ^{6}Li has only one single simultaneously trappable spin state such that depolarizing events, which are caused by emission and reabsorption of randomly polarized photons, more likely lead to trap loss. The lifetime of the atoms inside the magnetic trap was found to be half of that inside the MOTs, which we attributed to the different depths of both traps. The lifetime in the magnetic trap in the science chamber was measured to ~ 26 s.

- The **magnetic transport** of the atoms to the science chamber allows for a transfer efficiency of $\sim 15\%$. We believe that it is limited by atom losses on the walls of the narrow collimation tubes, through which the transport is directed. In addition, the long transport time of ~ 3 s leads to significant loss due to collisions with the background atoms. New power supplies have recently been installed in our setup to allow for tighter confinement of the atoms during the transport, which will decrease the losses on the walls as well as the total transport time.

Chapter 4

Photoassociation of heteronuclear ^{6}Li^{40}K molecules

In this chapter we investigate, both experimentally and theoretically, the formation of weakly bound, electronically excited, heteronuclear ^{6}Li^{40}K* molecules by single-photon photoassociation in a magneto-optical trap. We performed trap loss spectroscopy within a range of 325 GHz below the Li($2S_{1/2}$)+K($4P_{3/2}$) and Li($2S_{1/2}$)+K($4P_{1/2}$) asymptotic states and observed more than 60 resonances, which we identify as rovibrational levels of 7 of 8 attractive long-range molecular potentials. The long-range dispersion coefficients and rotational constants are derived. We find large molecule formation rates of up to $\sim 3.5 \times 10^7 \text{s}^{-1}$, which are shown to be comparable to those for homonuclear ^{40}K$_2^*$. Using a theoretical model we infer decay rates to the deeply bound electronic ground-state vibrational level $X^1\Sigma^+(v'=3)$ of $\sim 5 \times 10^4 \text{s}^{-1}$. Our results pave the way for the production of ultracold bosonic ground-state ^{6}Li^{40}K molecules which exhibit a large intrinsic permanent electric dipole moment. In this chapter, we also present the novel results obtained from photoassociation spectroscopy of ^{40}K$_2^*$ and compare them to prior studies with ^{39}K$_2^*$.

4.1 Introduction

4.1.1 Principle of photoassociation

In the process of photoassociation (PA), two colliding atoms (A and B) absorb a resonant photon (γ) which excites them to a bound molecular state:

$$A + B + \gamma \to (AB)^*. \tag{4.1}$$

This process, which can be considered as a light-induced chemical reaction, is schematically depicted in Fig. 4.1. Typically the energy of the photon is smaller than the energy E_{at} of the atomic transition of one of the colliding atoms. Once excited, the molecule will soon decay, either into a pair of free atoms or into a bound electronic ground-state molecule:

$$(AB)^* \to A + B + \gamma_{\text{d}} \qquad (4.2)$$

$$(AB)^* \to AB + \gamma_{\text{d}}. \qquad (4.3)$$

For the PA process to happen, the atoms need to approach each other sufficiently close. The probability for the PA transition is determined by the overlap of the initial wave function of the colliding atoms and that of the bound molecule. This so-called "Franck-Condon" overlap is typically largest at the internuclear separation R_{v+} for which the excited bound state has its classical outer turning point (see Fig. 4.1). Thus, the PA process most likely occurs at the separation R_{v+}.

Photoassociation can occur between atoms of the same species (homonuclear PA) or between atoms of different species (heteronuclear PA). The rates at which PA can occur in an atomic gas is typically smaller for the heteronuclear case due to the different range of the excited-state potentials. Whereas two identical atoms in their first excited state interact via the resonant dipole interaction at long range (with potential $V(R) \propto -C_3/R^3$, R denoting the internuclear separation and C_3 a constant), two atoms of different species interact via the van der Waals interaction ($V(R) \propto -C_6/R^6$, with a constant C_6), leading, for the heteronuclear case, to molecule formation at much shorter distances. Now, in a gas of atoms with a given density the probability to find two atoms at a distance R is approximately proportional to $(1 - e^{-4\pi nR^3/3})$, where n denotes the atom density, supposed to be homogeneous [1]. Consequently, fewer atom pairs with short internuclear distances are available, and the PA rate is smaller for the heteronuclear case. In this chapter we report on both homonuclear $^{40}K_2^*$ and heteronuclear $^6Li^{40}K^*$ photoassociation.

Photoassociation can be used as a spectroscopy technique which allows one to probe the states of the electronically excited molecular potential and, when the decay process is stimulated coherently with a second laser beam, also the states of the electronic ground-state potential. It has been proposed by Thorsheim et al [114] to apply PA

---

[1]This can be seen as follows: the probability for each atom to be located inside a sphere of volume $v = 4\pi R^3/3$ around a given atom is $p = nv/N$, where N denotes the total number of atoms. Thus, the probability to find at least one of the N atoms inside this sphere is $P = 1 - (1-p)^N = 1 - e^{N\ln(1-nv/N)} \approx 1 - e^{-nv} = 1 - e^{-4\pi nR^3/3}$. The probability to find two atoms at a distance R is then proportional to P.

4.1. Introduction

to ultracold atoms (which have small thermal energies E_{th}, see Fig. 4.1) in order to achieve high precision spectroscopic data for weakly bound molecules. Many more applications are offered when combining PA and ultracold atoms, which we will briefly describe in the subsequent section.

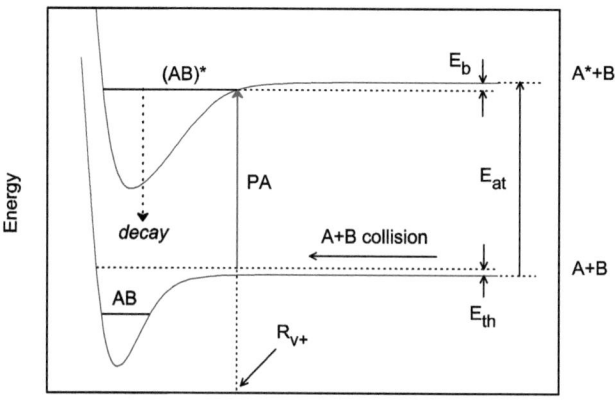

Figure 4.1: Schematic representation of the PA process. Two colliding atoms (A and B) are transferred into a bound state of the excited molecular potential by a photon of the PA laser, which is resonant with this transition. The curves represent the potential energy of the atoms as a function of their internuclear separation R. The PA process most likely occurs at the classical outer turning point R_{v+} of the bound excited state. E_{th}, E_{at}, E_b denote the thermal energy of the colliding atoms, the energy of the atomic transition of atom A and the binding energy of the molecule, respectively (not to scale). Once created, the excited bound molecule will decay, either to a bound state of the ground-state molecular potential or into a pair of free atoms.

4.1.2 Applications of ultracold photoassociation

There is an extensive literature describing ultracold photoassociation and its applications. We give a brief introduction here, for a more detailed review, we refer the reader to the articles [170, 171, 172, 173].

One of the principle applications of ultracold PA is molecular spectroscopy. Conventionally, molecular spectroscopy is performed with natural bound ground-state molecules formed in a high-pressure gas of atoms, typically above room temperature. These molecules are excited with a probe laser and frequencies for bound-bound transitions are determined. Ultracold photoassociation spectroscopy (UPAS) differs from this conventional molecular spectroscopy in several aspects associated with the low temperature. First, Doppler shifts due to the center-of-mass motion of the colliding atom pairs (along the direction of the PA laser) are reduced. Second, the thermal broadening due to the spread in the collision energies is reduced ($\sim$ 20 MHz for a 1 mK cold atom cloud). UPAS thus achieves high resolution for free-bound transitions. Third, PA allows the measurement of the binding energy of the molecules (it is simply given by the detuning of the PA laser from the atomic resonance), which is inaccessible for conventional spectroscopy. Due to the low temperatures, binding energies can thus be determined with a high precision. Fourth, rotational barriers prevent the low-energy atoms to approach each other when they are colliding in a higher-order partial wave. Thus only atoms colliding with very low angular momenta ($l = 0, 1$ for ^{6}Li-^{40}K collisions at $\sim$ 1 mK) can approach each other sufficiently close to form an excited bound molecule. Therefore, UPAS yields spectra with resonances from the lowest rotational levels only, which eliminates the need to extrapolate when determining a rotationless potential. Besides, UPAS allows probing of a larger number of excited molecular potentials, since the selection rules are less restrictive than for bound-bound transitions, because the initial atomic collision often possesses a mixture of many different molecular symmetries. Finally, since the free-bound transitions occur at large internuclear distance, UPAS is particularly powerful to probe long-range, weakly bound molecular states, which is difficult for conventional molecular spectroscopy, due to the reduced transition probabilities between tightly and weakly bound molecules. PA and conventional molecular spectroscopy can thus be considered as being complementary. Combining both methods allows the determination of precise molecular potential curves for the whole range of internuclear separations. In the results section we will present our recorded spectroscopic data for ^{40}K$_2^*$ and ^{6}Li^{40}K* and we will derive the binding energies of the most weakly bound states and the corresponding molecular potential curves at long-range.

Another application of UPAS is the precise determination of the lifetime of excited atomic states [174]. This application is based on the fact that the single atom dipole matrix element $d = \langle S|\mu|P\rangle$ governs both the atomic state lifetime and the form of the long-range part of the corresponding homonuclear excited molecular potential. Since the latter can be determined very precisely with UPAS, precise values for the atomic lifetimes can be derived. Such lifetime measurements provide a useful check of the more conventional atomic experiments, since the possible underlying systematic errors are different. In the results section of this chapter we will make use of our recorded PA spectra for homonuclear ^{40}K$_2^*$ molecules to determine the atomic lifetime of the $4P$-state of ^{40}K with a 0.3% precision.

PA can also be used to precisely determine the s-wave scattering length for binary

4.1. Introduction

ground-state collisions. This parameter is of central importance, since it contains all information inherent to the scattering event at ultracold temperatures, where only s-waves contribute to the elastic collisions. Scattering lengths sensitively depend on the precise details of the scattering potentials, which makes it difficult to predict them theoretically and thus assigns a great importance to experimental methods for their determination. The scattering length is defined as [175]

$$a = -\lim_{k \to 0} \left(\frac{\tan \eta_k^{l=0}}{k} \right), \qquad (4.4)$$

where $k = 2\pi/\lambda_{\mathrm{dB}}$ is the wave vector, λ_{dB} the de Broglie wavelength of the reduced mass particle and η_k is the s-wave phase shift induced by the scattering potential. The essential idea is to use PA spectroscopy to probe the phase of the ground-state wave function with great precision. Two different methods have been established to realize this. The first consists in the determination of the nodal positions of the ground-state wave function from the relative line contrasts of single-color PA spectra. The second is based on the determination of the binding energy of the least bound electronic ground state by two-color PA. The first method takes advantage of the fact that PA spectroscopy line intensities are sensitive to the amplitude of the ground-state wave function for each contributing partial wave in the region where PA occurs. Comparing the observed oscillations of the line intensities to quantum scattering calculations then allows the determination of the phase shift of the collision wave function and thus the scattering length. The second method employs two coherent laser beams, where one excites the colliding pairs to bound molecules and the other stimulatedly deexcites them to the last vibrational level of the electronic ground-state potential. The relative detuning of both lasers yields the binding energy E_{last} of that level. The scattering length can then be obtained by utilizing that in the ultracold limit, the scattering length is simply related to E_{last} via $E_{\mathrm{last}} = -\hbar^2/(2\mu a^2)$, where $\mu = m_A m_B/(m_A + m_B)$ is the reduced mass of the colliding particles [176]. The dependence of the scattering length on the binding energy of the least-bound state also demonstrates that it is very difficult to obtain an estimate for the scattering length from theoretical potential curves, as the depth of the molecular potentials is typically several orders of magnitude larger than E_{last}, such that a determination of E_{last} would require a very precise knowledge of the potential curve. Both the above described methods have been successfully applied in the past for Li [177, 178], Na [179], Rb [180], Cs [181], Yb [182]. Our recorded spectra which we will present in the results section might allow us to determine the scattering length using the first method. However, we concentrated the present work on the resonance assignment and the determination of the potential parameters and did not attempt to derive the scattering length.

The major long-term application of PA is the formation of ultracold molecules [98, 183]. As mentioned in chapter 1, molecules have a very complex internal level structure and are thus difficult to cool, since the laser cooling techniques, which work well for atoms, cannot be extended easily to molecules [108, 109]. PA allows one to circumvent this difficulty: photoassociating ultracold atoms results in molecules which are translationally cold as well. This fact is also utilized in the alternative technique of magnetoassociation via Feshbach resonances [184]. These two techniques represent indirect methods for creating ultracold molecules in contrast to direct methods, in which pre-existing (relatively hot) molecules are actively cooled [111, 113, 112]. In order to have sufficient time for experiments, one is interested in forming stable molecules. Multicolor PA allows the transfer of the associated atoms to bound ground-state molecules. In some cases, even single-photon PA followed by spontaneous decay to the lowest rovibrational level of the electronic ground-state potential can be high enough for an efficient production of stable molecules [84]. As discussed in chapter 1, the formation of ultracold molecules is particularly interesting for heteronuclear atoms, since molecules composed of two different atoms have a permanent electric dipole moment (as their charge distribution is asymmetric). The alkali dimer LiK has a very large dipole moment of 3.6 D [88] in its singlet electronic ground state and would thus be a good candidate for future studies with polar molecules.

4.1.3 Photoassociation of LiK* compared to other dimers

The first demonstration of PA of ultracold atoms has been realized in 1993 with sodium atoms confined in a magneto-optical trap [185]. Since then homonuclear PA has been demonstrated for all alkali atoms, *i.e.* Li$_2^*$ [177], Na$_2^*$ [185], K$_2^*$ [186], Rb$_2^*$ [187], Cs$_2^*$ [188], then H$_2^*$ [189], metastable He$_2^*$ [190], Ca$_2^*$ [191], Sr$_2^*$ [192] and Yb$_2^*$ [182]. Heteronuclear PA has been realized later, for the first time for the dimer ^{7}Li^{6}Li* [193] in 2002 and subsequently for RbCs* [194], KRb* [195], NaCs* [196], LiCs* [84] and YbRb* [197]. In this book, we report on the first realization of PA of LiK*.

It has been argued in the past, that the photoassociative creation of the dimer LiK* (as compared to the other heteronuclear dimers) would be more difficult to achieve due to its small reduced mass and C_6 coefficients. Wang and Stwalley [198] have proposed an estimate for the relative PA probabilities for the different heteronuclear alkali dimers based on the semiclassical reflection approximation [198]. They showed that for a given vibrational level number (counted from the dissociation limit) the PA probabilities for the different heteronuclear alkali dimers scale as $\mu^{9/4}(C_6)^{3/4}$, where μ is the reduced

4.1. Introduction

mass of the dimer. For the LiK* molecule this leads to small PA probabilities, of e.g. ~ 400 times less than for RbCs* and ~ 200 times less than for KRb*. Figure 4.2 summarizes the results of Ref. [198], showing the calculated relative Franck-Condon factors for the different heteronuclear molecules.

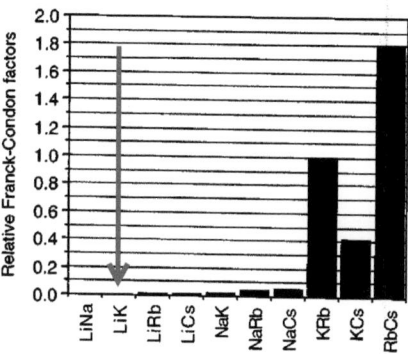

Figure 4.2: Relative values of the Franck-Condon factors for the different heteronuclear alkali dimers for PA at long-range, based on the reflection approximation [198]. The small value for LiK indicates that the photoassociative creation of LiK* molecules is expected to be much more difficult than for the other dimers. Figure taken from Ref. [198].

In the approximation made by Wang and Stwalley, the ground-state molecular potential is considered to have a negligible slope in the region where PA occurs. The authors justify this assumption by the fact that the ground-state molecular potential, which is also governed by the van der Waals interaction (with potential $V(R) \propto -C'_6/R^6$, and a constant C'_6), is of shorter range than the excited molecular potential due to a C'_6 coefficient which is typically about ten times smaller than the C_6 coefficient of the excited molecular potential. For homonuclear molecules, the difference in range of the ground-state and excited molecular potentials is orders of magnitude larger than for heteronuclear molecules, due to the strong resonant dipole interaction, such that for the case the assumption is well satisfied. For the heteronuclear case, the assumption is not sufficiently satisfied. The steepness of the excited molecular potential leads to molecule formation at small internuclear separations even for small binding energies where the ground-state potential has a non-negligible slope, as pointed out by Azizi et al. [199]. The authors of the latter publication computed the PA rates for several heteronuclear dimers (unfortunately not for LiK*) in a different way. Their results are based on full quantum mechanical numerical calculations of the Franck Condon overlaps considering PA to occur not only near the classical outer turning points of the excited molecular states. The authors obtained that indeed PA of lighter heteronuclear dimers is less probable, however, the difference in probability with respect to the heavier dimers is a factor of ~ 4 rather than ~ 100. Due to this discrepancy, we did not know which PA rates to expect in our experiment. In their work Azizi et al. [199] further compare the PA rates for heteronuclear molecules with those for homonuclear Cs_2^* molecules for given PA detunings, showing that the photoassociative formation of heteronuclear molecules for a given atomic density is about 10-30 times less probable due to the smaller distances at which PA occurs. This hinted that it would be difficult to observe PA of LiK*.

In our experiment, we observe PA rates for LiK* of the order of $\sim 3.5 \times 10^7 s^{-1}$. Given the above theoretical predictions, these rates are surprisingly large, since they are similar to those observed in a comparable experiment with RbCs* [194] and larger than those observed for other heteronuclear PA experiments [194, 195, 196, 84, 197]. This observation indicates the failure of the reflection approximation for the case of heteronuclear molecules and confirms the arguments of Ref. [199]. In our experiment we further observe that the PA rates for LiK* are comparable to those found for homonuclear K_2^*. The large PA rates for LiK* encourage future experiments aiming to create LiK ground-state molecules. In particular, for excited heteronuclear molecules the transition probabilities for spontaneous decay to ground-state molecules is larger than for homonuclear molecules due to the similar shape of the excited and ground-state potentials for the heteronuclear case [199]. Therefore high production rates for ground-state molecules can be expected.

4.1.4 Detection techniques for photoassociation

There are many techniques for observing photoassociation. One of the most common methods consists of a measurement of trap loss, where the loss in atom number which is induced by the molecule formation is recorded. The loss occurs since most of the excited molecules decay into hot atoms of ground-state molecules which are both not trapped by the atom trap. This method was used in the first demonstration of PA of Rb_2^* [187]. It is particularly convenient to apply it in a MOT, where the atom number can be continuously recorded via the fluorescence emitted by the atoms. When the frequency of the PA light is scanned the formation of bound excited states then appears as a dip in the fluorescence signal. Another very common method consists in making an additional excitation to an ionizing state—either from the electronically excited state or from one of subsequently populated states of the electronic ground state—and to measure the ion rate with an ion detector. This method was used in the first demonstration of

4.1. Introduction

PA of Na$_2^*$ [185]. Another technique, which is adapted to detect predissociating excited molecular levels is fragmentation spectroscopy [200], which we will not further consider here.

Each technique has its advantages and disadvantages. The ionization technique is very sensitive since it has nearly zero background signal. When recording a PA spectrum the PA laser thus can be scanned very fast. In addition even free-bound transitions with very low probabilities can be probed, allowing one to record PA spectra for a large range of detunings. Furthermore, a higher spectral resolution can be obtained as the PA laser can be used at lower intensities decreasing the effect of power broadening of the resonances. However, the ionization technique requires the installation of an ionization laser and an ion detector. Furthermore the measurement of transition probabilities is difficult, since it requires the knowledge of the probabilities for the spontaneous decay and the ionization, which is not always available.

The trap loss technique has the advantage that its implementation takes less effort and that in particular it gives access to the formation rates for the excited molecules, since only the molecular transition is probed. As we will show in the results section, the measurement of the formation rates is crucial for the identification of the observed resonances. On the other hand, the trap loss technique has the following disadvantages. In our experiment, PA is performed with atoms trapped in a MOT for which the measurement of trap loss requires the PA laser to be scanned slowly since the response time of the MOT is rather slow ($\sim 1\,\mathrm{s}$). In addition, only resonances with a contrast larger than the (averaged) fluctuations of the MOT fluorescence can be detected which requires a stable MOT operation and to record many data for averaging. Besides, only excited molecules which decay into a bound molecule or dissociate to a pair of atoms with an energy larger than the depth of the trap can be detected. Recaptured atoms do not contribute to the trap loss signal. The probability for recapture of dissociating atoms can be decreased by lowering the depth of the trap. In our experiment we indeed observed, that the PA-induced trap loss for both ^{40}K$_2^*$ and ^{6}Li^{40}K* could be significantly increased for some resonances when the capture velocity was lowered (by decreasing the detuning of the trapping light). Another possibility to circumvent the recapture problem is the detection of the PA-induced heating of the atom cloud rather than the loss of atoms, as has been implemented by Leonard et al. [201] with metastable helium atoms confined in a magnetic trap. This technique is, however, not applicable to a MOT, since the atoms are continuously cooled very efficiently.

In our experiment we applied the trap loss technique. Most of its disadvantages could be overcome by carefully adjusting the experimental parameters. Furthermore, the application of this technique allows us to directly compare the formation rates for

^{6}Li^{40}K* molecules with those for ^{40}K$_2^*$ molecules.

4.1.5 Molecular potentials

In the following we give a short overview of the different possible interactions between two alkali atoms and the resulting molecular potentials, which are relevant for the analysis of our spectroscopic data.

For a diatomic molecule a molecular potential curve represents the total electrostatic energy of the molecule as a function of the distance of the two atoms. The notion of a molecular potential is justified by the fact that the electrons of the atoms are much lighter than the nuclei and thus can immediately adjust to the much slower motion of the nuclei. This is the basis of the well-known Born-Oppenheimer approximation which allows the calculation of the energy assuming that the wave function of the molecule separates into a product of a nuclear and an electronic part. The potential curves which are obtained when only electrostatic interactions and electron exchange are considered, are in the following referred to as the Born-Oppenheimer potentials. Additional interactions exist, such as the spin-orbit and hyperfine interactions within each atom, but shall be neglected for the time being. For small internuclear distances, where the electrostatic and exchange interactions are the dominant interactions, this neglect is typically justified.

When the two atoms are in different electronic states the molecular potentials take different forms. Figure 4.3 (a) shows the Born-Oppenheimer potentials dissociating to the three lowest electronic asymptotes $2S+4S$, $2S+4P$ and $2P+4S$ of the LiK molecule. They have been calculated as described in detail in Ref. [88]. The potentials are labeled by quantum numbers indicating the symmetry of the electronic wave functions. Those can be inferred from the atomic properties at the dissociation limit. When rotation, spin-orbit and hyperfine interactions are neglected, the total electron spin $\boldsymbol{S} = \boldsymbol{s}_A + \boldsymbol{s}_B$ is conserved. For alkali atoms it is $s = 1/2$, resulting in either $S = 1$ (triplet states) or $S = 0$ (singlet states). In contrast, the total angular momentum $\boldsymbol{L} = \boldsymbol{l}_A + \boldsymbol{l}_B$ of the atoms is not conserved, since the molecule is not spherically symmetric. It is, however, cylindrically symmetric around the internuclear axis, such that the projection of $\boldsymbol{L}$ onto this axis, $\Lambda = \lambda_A + \lambda_B$, is conserved, where the λ_i are the respective projections of the l_i. For example, for potentials dissociating to the $S+P$ asymptote, it is either $\Lambda = 0$ or $|\Lambda| = 1$. The molecular potentials are labeled by $^{2S+1}|\Lambda|$, where Λ is represented as $\Sigma, \Pi, \Delta, ...$ for $|\Lambda| = 0, 1, 2, ...$, respectively. The superscript $\pm$ indicates a reflection symmetry of the spatial component of the electronic wave function through a plane containing the internuclear axis. Homonuclear molecules are labeled with an additional

4.1. Introduction

subscript g/u (for gerade/ungerade), which indicates the parity with respect to an inversion of all electrons through the center of charge. In addition the potentials are numbered. Potentials with the same symmetry (which do thus not cross) are numbered with increasing energy as $\mathcal{N} = 1, 2, 3, \ldots$. The two electronic ground-state potentials take the "numbers" X and a instead of 1 for historical reasons.

Long-range potential energy curves

The molecular states which are probed by PA spectroscopy are typically of long-range, and their properties are mainly determined by the long-range part of the molecular potentials. At large internuclear distances, the potential energy of a diatomic molecule can be approximately calculated from the properties of the separated atoms alone using perturbation theory. When the spatial overlap between the charge distributions of the two atoms can be neglected, the potential energy can be written as a sum of inverse powers of the internuclear distance R:

$$V(R) = D - \sum_n \frac{C_n}{R^n}, \qquad (4.5)$$

where D is the dissociation energy of the molecule (*i.e.* approximately the depth of the potential). The constants C_n are called the dispersion coefficients. Depending on the interaction between the atoms, different terms in this expression are zero. When the interaction is of van der Waals type, the two leading order terms are $n = 6$ and $n = 8$. A ground-state or a singly excited heteronuclear atom pair (*e.g.* K-K, Li-K or Li-K*) interact via this interaction. For a singly excited homonuclear atom pair (*e.g.* K-K*) the interaction is of resonant dipole type and the two leading orders are $n = 3$ and $n = 6$. Due to the different molecular symmetries each dispersion coefficient can assume two values for a given asymptote, one for singlet and one for triplet states.

So far, we have neglected spin-orbit and hyperfine interactions. These interactions are much smaller than the electrostatic interaction at short range, but they become dominant for the large interatomic distances, at which PA typically occurs and thus need to be considered. They lead to a separation of the short-range molecular potential curves into several long-range potentials. For large internuclear distances the molecular electronic wave function is approximately given by a product of the single-atom wave functions. The molecular spin-orbit interaction is thus approximately given by the sum of the atomic spin-orbit interactions. Consequently, the total electronic angular momentum of the molecule is given by $\boldsymbol{j} = \boldsymbol{j}_\text{A} + \boldsymbol{j}_\text{B}$, where the $\boldsymbol{j}_\text{i} = \boldsymbol{l}_\text{i} + \boldsymbol{s}_\text{i}$ are the total angular momenta of the atoms, respectively. $\boldsymbol{j}$ is not conserved since the molecule is not spherically symmetric. Its cylindrical symmetry, however, results in a conservation

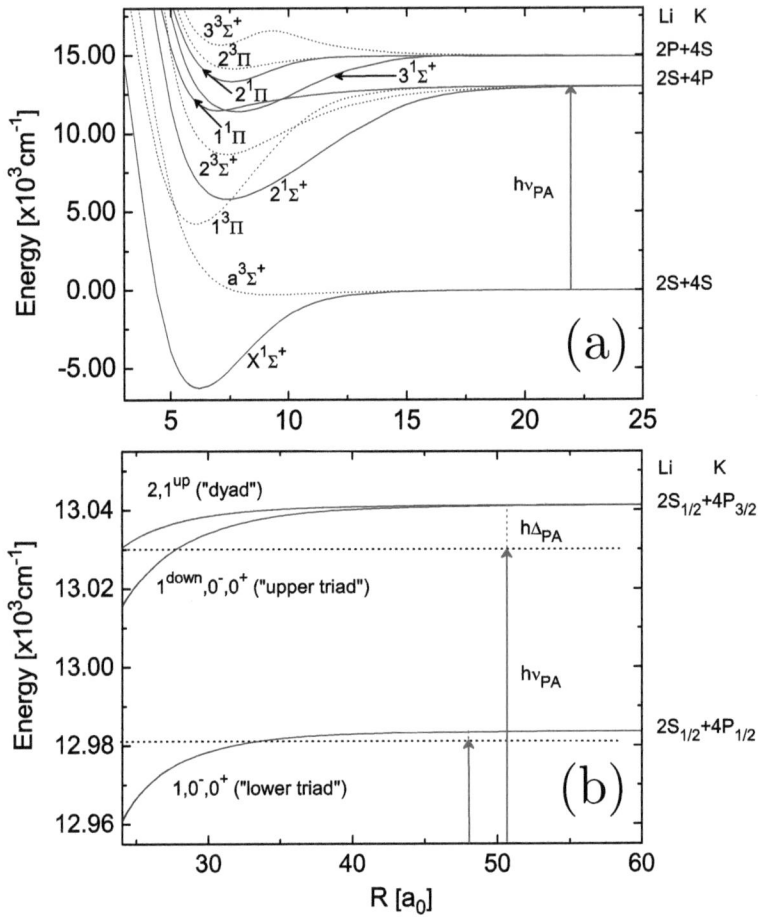

Figure 4.3: (a) Born-Oppenheimer potentials of the LiK molecule for short interatomic separations R (in units of the Bohr radius a_0). The potentials are labeled by $N^{2S+1}|\Lambda|^{(\pm)}$. (b) Long-range potential energy curves including spin-orbit interaction. The potentials are labeled by $\Omega^{(\pm)}$ and further designated by us by the superscripts up/down and the classification into the potential groups "lower triad","upper triad" and "dyad" for unambiguous distinction. For large R the potential curves are approximated by $V(R) = D - C_6/R^6$. Potentials with the same C_6 coefficient are thus hardly distinguishable in the figure. Each of the curves in figure (b) approaches one of the curves in figure (a) according to Tab. 4.1. The vertical arrows represent the minimum energy delivered by the PA laser in our experiments.

4.1. Introduction

of the projection of j onto the interatomic axis, which is denoted as Ω.

Figure 4.3 (b) shows the long-range potentials dissociating to the asymptotes $2S_{1/2}+4P_{1/2,3/2}$ of the LiK* molecule, including the spin-orbit interaction. At short internuclear separations each of these potentials approaches one of the short-range potential curves of Fig. 4.3 (a) according to Tab. 4.1 [202]. The long-range potentials are labeled by $\Omega^{(\pm)}$. The superscripts $\pm$ indicate a reflection symmetry of the total electron wave function through a plane containing the internuclear axis. The superscripts apply only to the states with $\Omega = 0$. Since the labeling is not unambiguous we here further label the potentials with the superscripts up/down and classify them into groups for distinction. The potentials $1, 0^+, 0^-$ dissociating to the $2S_{1/2}+4P_{1/2}$ asymptote are referred to as the lower triad, the potentials $1^{\text{down}}, 0^+, 0^-$ dissociating to the $2S_{1/2}+4P_{3/2}$ asymptote are referred to as the upper triad and the potentials $2, 1^{\text{up}}$ are referred to as the dyad. Not shown in Fig. 4.3 (b) are the long-range potentials which dissociate to the ground-state asymptote $2S_{1/2}+4S_{1/2}$ and the excited-state asymptotes $2P_{1/2,3/2}+4S_{1/2}$. The former is approached by three attractive long-range potentials $1, 0^+, 0^-$. The latter are approached by eight long-range potentials, which are all repulsive [198].

Short range	Long range	Asymptote
$1^1\Pi$	1^{up}	$2S_{1/2}+4P_{3/2}$
$2^3\Sigma^+$	$1^{\text{down}}, 0^-$	
$2^1\Sigma^+$	0^+	$2S_{1/2}+4P_{1/2}$
$1^3\Pi$	$2, 0^+$	$2S_{1/2}+4P_{3/2}$
	$0^-, 1$	$2S_{1/2}+4P_{1/2}$

Table 4.1: Correlation table for the molecular potentials dissociating to the asymptotes $2S_{1/2}+4P_{1/2,3/2}$ of LiK*.

The potentials shown in Fig. 4.3 (b) still have the asymptotic behavior given by Eq. (4.5). The values of the dispersion coefficients are, however affected by the spin-orbit interaction. For the case of LiK, the dispersion coefficients can, for the relevant $2S_{1/2}+4P_{1/2,3/2}$ asymptotes, only assume three (rather than eight) different values due to the small atomic fine structure of the Li atom [203]. Potentials with the same C_6 coefficient are thus hardly distinguishable in Fig. 4.3 (b). The dispersion coefficients C_6 and C_8 of the LiK molecule have been calculated theoretically [203, 166, 167] and are listed in Tab. 4.2 together with the experimentally determined values obtained in this work, those of K_2 can be found in [204, 205].

Asymptote	Potential	C_6 [$\times 10^3$]	C_8 [$\times 10^5$]	Reference
$2S_{1/2}+4S_{1/2}$	$1, 0^+, 0^-$	2.27	—	[203]
		2.55	1.75	[166]
		2.29	1.92	[167]
$2S_{1/2}+4P_{1/2}$	$1, 0^+, 0^-$	15.42	—	[203]
		13.83	34.19	[166]
		15.81*	—	[167]
	0^+	12.86 ± 0.66	—	this work
$2S_{1/2}+4P_{3/2}$	$2, 1^{\text{up}}$	9.52	—	[203]
		9.80	0.44	[166]
		9.34*	—	[167]
	2	9.17 ± 0.94	—	this work
	1^{up}	9.24 ± 0.96	—	this work
	$1^{\text{down}}, 0^+, 0^-$	22.00	—	[203]
		25.50	11.20	[166]
		22.17*	—	[167]
	1^{down}	25.22 ± 0.60	—	this work
	0^+	25.45 ± 0.72	—	this work
	0^-	24.31 ± 1.71	—	this work

Table 4.2: Theoretical predictions and experimentally determined values obtained in this work for the long-range dispersion coefficients for the lowest asymptotes of LiK given in atomic units (a.u.), where 1 a.u.=1 $E_h a_0^6$ for C_6 and 1 a.u.=1 $E_h a_0^8$ for C_8, with the Hartree energy $E_h = 4.35974 \times 10^{-18}$ J and the Bohr radius $a_0 = 5.29177 \times 10^{-11}$ m. The experimental values agree best with the predictions of Ref. [166]. In Ref. [167] the dispersion coefficients have been calculated for the absence of spin-orbit interactions only. Including the spin-orbit interaction according to the method described in Ref. [203] yields the values given in the table (marked with an asterisk) [206].

Mechanical rotation of the molecule

So far, we have neglected the mechanical rotation of the molecule, which we now consider. Neglecting the internal electronic structure of the two atoms, their relative motion is described by the Schrödinger equation

$$\left[-\frac{\hbar^2}{2\mu}\nabla^2 + V(\boldsymbol{R})\right]\Psi(\boldsymbol{R}) = E\Psi(\boldsymbol{R}), \tag{4.6}$$

where E is the atoms' relative kinetic energy, $V(\boldsymbol{R})$ their interaction potential and $\boldsymbol{R}$ the vector connecting both atoms. Since V is spherically symmetric and approaches zero at infinity, the wave function can be separated in radial and angular parts

$$\Psi(\boldsymbol{R}) = \sum_{l=0}^{\infty} \sum_{m=-l}^{m=l} \frac{u_l(R)}{r} Y_{l,m}(\vartheta, \varphi), \tag{4.7}$$

4.1. Introduction

with the spherical harmonic functions $Y_{l,m}(\vartheta,\varphi)$. The Schrödinger equation for the radial part becomes

$$\left[-\frac{\hbar^2}{2\mu}\frac{\partial^2}{\partial R^2} + V(R) + \frac{\hbar^2 l(l+1)}{2\mu R^2}\right] u_l(R) = E u_l(R), \tag{4.8}$$

where l is the quantum number of the angular momentum $\boldsymbol{l}$ associated with the atom's relative motion. The second and third terms inside the square brackets on the left hand side can be considered an effective potential $V_{\text{eff}}(R) = V(R) + \hbar^2 l(l+1)/(2\mu R^2)$. Thus, the rotation of the molecule changes its potential energy by the simple term $\hbar^2 l(l+1)/(2\mu R^2)$. For diatomic molecules $\boldsymbol{l}$ is perpendicular to the internuclear axis.

Strictly speaking, l is not a good quantum number since other angular momenta are present in the molecule, which may cause the molecular axis to nutate and precess. However, the absolute value J of the total angular momentum $\boldsymbol{J} = \boldsymbol{l} + \boldsymbol{L} + \boldsymbol{S}$ is always a good quantum number. The rotational energy levels for a given vibrational level of the molecular potential are given by $E_{\text{rot}} = B_v X(X+1) + ...$, where B_v is the rotational constant and X an angular momentum quantum number, which is determined by the way how the different angular momenta of the atoms couple. That in general is determined by the dominant interaction, which, typically as a function of the distance R, is classified by Hund's coupling cases [207]. In general, in the limit of very large R the rotational energy is small compared to the electrostatic and spin-orbit interaction and the valid case is Hund's case (c). In this case the angular momentum quantum number which determines the rotational energy is $X = J$. Neglecting centrifugal distortion, the rotational energy in Hund's case (c) is more precisely given by [207, 171]

$$E_{\text{rot}} = B_v[J(J+1) - \Omega^2], \quad \text{with } J = \Omega, \Omega+1, \Omega+2, ... \tag{4.9}$$

(the relation $J \geq \Omega$ results from the definition of Ω). Neighboring rotational energy levels will thus be spaced by $\Delta E_{\text{rot}} = E_{\text{rot}}(J+1) - E_{\text{rot}}(J) = 2B_v, 4B_v, 6B_v, ...$ This formula will be of importance for the assignment of our heteronuclear PA spectra in the results section.

4.1.6 Selection rules

The knowledge of the selection rules for the electric dipole transitions induced by the PA laser is another requirement for the assignment of our heteronuclear PA spectra. On the one hand, the selection rules allow us to infer which excited molecular potentials are accessible via PA and on the other hand how many rotational lines per vibrational level are expected. We thus briefly mention the relevant selection rules. Since the photon carries at most one unit of angular momentum, the total angular momentum of the atom pair can change by at most one in a PA transition: $\Delta J = 0, \pm 1$. In Hund's case (c) this selection rule implies $\Delta \Omega = 0, \pm 1$ [207, 171]. Since the photon acts on the electronic orbital degrees of freedom, it cannot change the electron spin of the molecule: $\Delta S = 0$. In Hund's case (c), the photon couples states with identical $+/-$-symmetry. Besides, for homonuclear molecules, the photon couples g to u states and vice versa and thus transitions between states of identical g/u-symmetry are forbidden.

4.1.7 Rotational barriers for ultracold ground-state collisions

Earlier we mentioned that in ultracold PA only slowly rotating molecules can be created. This is a consequence of the repulsive interaction between two ground-state atoms at long range which results when they collide in a partial wave of higher-order ($l > 0$). This repulsive interaction creates the so-called rotational barriers which can prevent the atoms to approach each other sufficiently close for PA to occur. In the following we calculate the height of the rotational barriers for different partial waves for two colliding ground-state atoms for the case of K-K and Li-K collisions. From the obtained heights we derive the highest allowed partial wave order for the given collisional energies, $i.e.$, the highest allowed value for the angular momentum quantum number l in order to predict the number of expected rotational levels in our recorded spectra.

For two colliding ground-state atoms, the interatomic potential $V(R)$ at long range can be approximated by $V(R) = D - C_6'/R^6$, so the effective potential for the colliding atoms takes the approximate form (see Eq. (4.8))

$$V_{\text{eff}}(R) = D - \frac{C_6'}{R^6} + \frac{\hbar^2 l(l+1)}{2\mu R^2}. \quad (4.10)$$

This potential is plotted in Fig. 4.4 for different values of l for the relevant collision partners. The position R_{rb} and height $E_{\text{rb}} = V(R_{\text{rb}}) - D$ of the rotational barriers as a function of l are given by

$$R_{\text{rb}} = \left(\frac{6\mu C_6'}{\hbar^2 l(l+1)} \right)^{1/4}, \quad (4.11)$$

$$E_{\text{rb}} = \left(\frac{\hbar^2 l(l+1)}{3\mu (2C_6')^{1/3}} \right)^{3/2}. \quad (4.12)$$

For the lowest-order partial waves the position of the rotational barriers lies at large internuclear distances $R_{\text{rb}} \sim 100\, a_0$. The height of the rotational barriers is for K-K

4.1. Introduction

collisions $E^{KK}_{l=1} = 0.27\,\text{mK}$, $E^{KK}_{l=2} = 1.5\,\text{mK}$ and $E^{KK}_{l=3} = 4.1\,\text{mK}$ and for Li-K collisions $E^{LiK}_{l=1} = 2.6\,\text{mK}$, $E^{LiK}_{l=2} = 13.4\,\text{mK}$ and $E^{LiK}_{l=3} = 38.2\,\text{mK}$. For the collisional energies present in the MOTs, which have temperatures of $0.3\,\text{mK}$ (K-MOT) and $1.2\,\text{mK}$ (Li-MOT), only atoms colliding in partial waves with $l = 0, 1$ can thus reach small internuclear separations (with a reasonable probability) at which PA can occur. Collisions with higher l will be reflected from the potential at large internuclear distances unless tunneling occurs or the colliding atoms have a high velocity, which is, however, very unlikely due to the low temperatures and the large height and width of the rotational barriers for larger l. In our experiment, the fermionic nature of the atoms ^{40}K and ^{6}Li does not prevent the scattering to occur in both even and odd partial waves, since the atoms are trapped in a MOT and are thus not polarized.

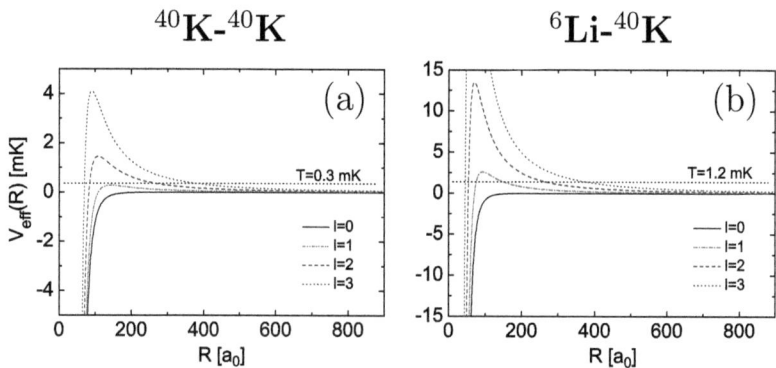

Figure 4.4: Effective long-range scattering potentials (Eq. (4.10), with D set to zero for convenience) for binary collisions involving (a) two ground-state ^{40}K atoms and (b) one ^{6}Li and one ^{40}K atom in their ground states for different l. At the collision energies of $\sim 0.3\,\text{mK}$ in the ^{40}K-MOT and $\sim 1.2\,\text{mK}$ in the ^{6}Li-^{40}K-MOT only $l=0$ (s-wave) and $l=1$ (p-wave) collisions reach short distances with significant probability, allowing photoassociation of the atoms. Higher angular momentum collisions are reflected from the larger rotational barriers at large internuclear distances.

We have seen that only atoms with an angular momentum $l = 0$ or $l = 1$ can reach sufficiently small internuclear distances for PA. We now specify the resulting rotational levels which can be addressed by PA for the case of LiK. In the initial collision, the value of the orbital angular momentum is $L' = 0$, since both atoms are in their electronic ground state [2]. The value of the total spin can take the values $S' = 0, 1$. Thus, the total angular momentum J' of the initial collision can be $J' = 0, 1, 2$. The selection

---
[2] L' is strictly speaking not a good quantum number, but it is reasonable to consider it to be one for the current purpose.

rule $\Delta J = 0, \pm 1$ for the PA transition now yields that the total angular momentum of the excited molecule can take the values $J = 0, 1, 2, 3$.

We can be even more specific. When the excited molecule has a certain symmetry Ω, the values of J are further restricted. We have already seen in Eq. (4.9) that J has an Ω-dependent lower bound, *i.e.*, $J \geq \Omega$. In the same way, J also has an Ω-dependent upper bound: in the Hund's case (c) representation, the initial collision has one of the symmetries 0^+ (for $X^1\Sigma^+$), 0^- or 1 (for $a^3\Sigma^+$, see Fig. 4.3). Due to $l = 0, 1$, the total angular momentum of the initial collision can thus take the values $J' = 0, 1$ for the initial states with $\Omega' = 0$ and $J' = 0, 1, 2$ for the $\Omega' = 1$ state. Therefore, the excited molecule can take the maximum value $J = 3$ only, if the state of the initial collision is $\Omega' = 1, J' = 2$ (since $\Delta J \leq 1$). Whereas the states $\Omega = 2, J = 3$ and $\Omega = 1, J = 3$ can be reached from this state, the states with $\Omega = 0, J = 3$ cannot be reached, as this transition would require $\Delta\Omega = -1$ and $\Delta J = +1$ at the same time, which is contradicting. For the potentials with $\Omega = 0$ the maximum allowed value of J is $J = 2$, since $\Omega = 0, J = 2$ can be reached, *e.g.*, from the state $\Omega' = 0, J' = 1$. The accessible rotational quantum numbers J for the different excited molecular potentials are summarized in Tab. 4.3. The knowledge of the accessible rotational levels for each

Ω	J	ΔE_{rot}
2	2, 3	$6B_v$
1	1, 2, 3	$4B_v, 6B_v$
0	0, 1, 2	$2B_v, 4B_v$

Table 4.3: Accessible rotational quantum numbers J and observable rotational splittings $\Delta E_{\text{rot}} = E_{\text{rot}}(J+1) - E_{\text{rot}}(J)$ between neighboring rotational levels (see Eq. (4.9)) of the excited LiK* molecule for the different molecular potential symmetries Ω for PA at the given atom temperature.

potential will be indispensable for the assignment of our heteronuclear PA spectra in the results section.

4.1.8 The LeRoy-Bernstein formula

In order to experimentally determine an accurate potential energy curve, in principle the binding energies of all vibrational bound states have to be measured. In PA, very often only the most weakly bound vibrational levels of the excited molecular potentials can be probed. The incomplete data set can, nonetheless, be used to determine the long-range part of the potential curve, since the properties of the weakly bound molecular states are mainly determined by that part of the potential. In 1970 LeRoy, Bernstein and Stwalley [208, 209] derived a semiclassical formula, which relates the

4.1. Introduction

binding energy of the weakly bound states to the shape of the potential, given that the potential can be approximated by

$$V(R) = D - C_n/R^n \tag{4.13}$$

at long range (*i.e.*, that the higher-order terms in Eq. (4.5) and the exchange energy can be neglected). Using the Wentzel-Kramers-Brillouin (WKB) semiclassical method the authors showed that the binding energy $D - E_v$ of the v-th vibrational state (counted from the dissociation limit D) is given by

$$D - E_v = A_n(v_D - v)^{(2n/(n-2))}, \tag{4.14}$$

with

$$A_n = C_n \left[\frac{\pi(n-2)\hbar}{B(1/n + 1/2, 1/2)\sqrt{2\mu C_n}} \right]^{2n/(n-2)}, \tag{4.15}$$

where B denotes the Beta-function (with $B(5/6, 1/2) \approx 2.241$ for $n = 3$ and $B(2/3, 1/2) \approx 2.587$ for $n = 6$), μ is the reduced mass of the system, C_n the leading-order dispersion coefficient of the molecular potential with $n = 3$ for excited homonuclear and $n = 6$ for excited heteronuclear molecules. v_D is a constant between 0 and 1 which describes the effective vibrational quantum number of an imaginary state whose energy corresponds to the dissociation limit (the most weakly bound state having $v = 1$). Equation (4.14) allows the derivation of the leading-order dispersion coefficient C_n from the measured binding energies of the weakly bound states. Due to the assumption about the form of the molecular potential (Eq. (4.13)) only vibrational states with zero rotational energy $(J(J+1) - \Omega^2 = 0)$ may be considered. The derivation of the dispersion coefficient is most conveniently done by plotting the $1/n$-th power of the measured binding energies of the vibrational levels versus their vibrational quantum numbers v and fitting it to a straight line, whose slope gives access to C_n.

In the results section of this chapter we will make use of the LeRoy-Bernstein (LRB) formula in order to determine the leading order dispersion coefficient for the different potentials probed in our experiment. Since the LRB formula (Eq. (4.14)) is based on several assumptions its validity has to be carefully checked beforehand in order to obtain information about its reliability. Typically, the biggest restriction of the validity of the LRB formula comes from the assumption of the molecular potential to be of the form $V(R) = D - C_n/R^n$ (Eq. (4.13)). This assumption neglects the exchange energy and it neglects higher-order terms in the long-range potential function (Eq. (4.5)), both of which become important for shorter internuclear distances. An estimate for the smallest distance above which the contribution of the exchange energy to the total energy of the molecule is less than 10% is given by the modified LeRoy radius [210]

$$R_{\text{crit}}^{(1)} = 2\sqrt{3}\left(\langle z^2\rangle_A^{1/2} + \langle z^2\rangle_B^{1/2}\right), \qquad (4.16)$$

where $\langle z^2\rangle^{1/2}$ is the rms distance of the valence electron from its nucleus along the internuclear axis (for atoms A and B). In our case, we find for the different excited molecular potentials

$$R_{\text{crit}}^{(1)} = \begin{cases} 31\,a_0 & \text{for K}_2^* \ (|\Lambda|=0), \\ 22\,a_0 & \text{for K}_2^* \ (|\Lambda|=1), \\ 28\,a_0 & \text{for LiK}^* \ (|\Lambda|=0), \\ 20\,a_0 & \text{for LiK}^* \ (|\Lambda|=1). \end{cases} \qquad (4.17)$$

An estimate for the smallest distance above which the contribution of the higher-order terms in the long-range potential function to the molecule's total energy is less than 10% is given by

$$0.1\frac{C_n}{\left(R_{\text{crit}}^{(2)}\right)^n} = \frac{C_m}{\left(R_{\text{crit}}^{(2)}\right)^m}, \qquad (4.18)$$

where m refers to the second order term in the expansion of Eq. (4.5). Using the dispersion coefficients given in Ref. [204] (see Tab. 4.2) one obtains

$$R_{\text{crit}}^{(2)} = \begin{cases} 20\,a_0 & \text{for K}_2^*, \\ 50\,a_0 & \text{for LiK}^* \text{ (lower triad)}, \\ 21\,a_0 & \text{for LiK}^* \text{ (upper triad)}, \\ 7\,a_0 & \text{for LiK}^* \text{ (dyad)}, \end{cases} \qquad (4.19)$$

where for K_2^* only the molecular potentials, for which we observe resonances in the experiment, are considered (which have the same dispersion coefficients at short internuclear distances where the spin-orbit interaction can be neglected). In order to judge, whether the assumption (4.13) of the LRB formula is fulfilled, the calculated critical separations $R_{\text{crit}}^{(1)}$ and $R_{\text{crit}}^{(2)}$ have to be compared with the minimum distance R_{min} at which PA occurs in the experiment. An estimate for R_{min} can be obtained from Eq. (4.13), assuming that PA occurs at the classical outer turning point, which scales with the PA detuning as $R_{v+} = [-C_n/(\hbar\Delta_{\text{PA}})]^{1/n}$ with $n=3$ or $n=6$, respec-

4.1. Introduction

tively. In our experiment, we have created homonuclear K_2^* and heteronuclear LiK^* molecules for PA detunings $|\Delta_{PA}|$ up to 313 GHz with respect to the corresponding $S+P_{3/2}$ dissociation limit. Consequently

$$R_{\min} = \begin{cases} 65\, a_0 & \text{for } K_2^*, \\ 24\, a_0 & \text{for } LiK^*, \end{cases} \quad (4.20)$$

where the smallest C_n coefficients of the different excited molecular potentials were considered. One can immediately see that the heteronuclear molecules formed in our experiment have a much smaller size than the homonuclear molecules for a given detuning. For the case of K_2^* it is $R_{\min} > R_{\text{crit}}^{(1)}, R_{\text{crit}}^{(2)}$ such that the assumption (4.13) of the LRB formula is well fulfilled. We therefore expect the LRB formula to give precise estimates for the potential's leading order coefficient when applying it to the recorded data. For the case of LiK^*, due to the much shorter distances at which PA occurs, the considered assumption is at the limit of being fulfilled. We thus expect the LRB formula to yield less precise estimates for the leading order dispersion coefficients.

Another assumption of the LRB formula is the validity of the WKB method, which does not hold for levels very close to the dissociation limit [211, 212]. In addition, the LRB formula does not consider hyperfine interactions and retardation effects, which are most pronounced for bound states with a very long range. Besides, since small binding energies can be measured with less relative accuracy (due to the uncertainties of the determination of the PA detuning and of the exact line positions which arise from unresolved hyperfine substructure and light shifts), we will avoid taking the most weakly bound vibrational levels into account when applying the LRB formula.

4.1.9 Previous work on LiK

Compared to the other heteronuclear alkali dimers, LiK has been studied relatively little. The first high-resolution spectroscopic results were obtained in 1984 for the $X^1\Sigma^+$-$1(B)^1\Pi$ transition [213]. Preliminary potential curves for the two involved potentials could be derived. Since then precise investigations of the ground-state potentials $X^1\Sigma^+$ and $a^3\Sigma^+$ [214], as well as for the excited potentials $1(B)^1\Pi$ [215, 216], $3(C)^1\Sigma^+$ [215, 216], $2(D)^1\Pi$ [217] and $4^1\Pi$ [218], which are all accessible form the ground-state potential $X^1\Sigma^+$, have been made. Only transitions to deeply bound excited molecular levels could be studied, since small Franck-Condon overlaps prohibit the excitation of weakly bound states from the deeply bound ground-state molecules available in the experiment.

Levels in the neighborhood of our PA spectra have been recorded for the $1(B)^1\Pi$

potential up to 390 ± 150 GHz below the $2S_{1/2}+4P_{3/2}$ asymptote [215]. The authors recorded the lowest 29 vibrational lines, excited from the $X^1\Sigma^+(v=3, J=14)$-level. The energies of the observed rotationally excited vibrational lines were extrapolated to $J=0$ in order to obtain a potential energy curve for the rotationless molecule. The bound-bound transition frequencies have been determined with an uncertainty of ± 3 GHz. The potential's dissociation energy was determined from the data via an extrapolation based on the LRB formula using the energies of the states with smallest binding energy. The dissociation energy could be determined with an uncertainty of $\pm h \times 150$ GHz. At large interatomic separations the potential $1(B)^1\Pi$ turns into the 1^{up} potential, in which we observed the five most weakly bound vibrational states up to a PA detuning of 313 GHz. It is likely that the most deeply bound level observed in our experiment corresponds to the most weakly bound level observed in Ref. [215]. Combining both spectroscopic data would thus allow the derivation of a precise determination of the $1(B)^1\Pi$ potential curve and its dissociation energy, similar to what has been done for LiCs* [219]. Since the spectroscopic data of Ref. [215] was recorded for the isotopomer ^{7}Li^{39}K and our PA data for the isotopomer ^{6}Li^{40}K, isotopic corrections will have to be applied [91].

Ultracold ^{6}Li^{40}K molecules in their electronic ground-state potential have recently been studied via Feshbach spectroscopy [90]. Loss features of 13 Feshbach resonances were observed and assigned based on an asymptotic bound state model and a full coupled channels calculation. The heteronuclear singlet and triplet s-wave scattering lengths have been derived from the spectra and found to be $a_s = 52.1\, a_0$ and $a_t = 63.5\, a_0$. The measured resonance positions were subsequently used to refine the electronic ground-state molecular potentials [91]. Weakly bound ^{6}Li^{40}K molecules in their electronic ground-state potential have been created for the first time in 2009 [93] by a magnetic field sweep across an interspecies s-wave Feshbach resonance.

4.2 Experimental results

In the previous section we have introduced the general background of ultracold photoassociation and spectroscopic concepts, which provides us with the necessary knowledge to understand the PA spectra recorded in our experiments. In this section we present first the experimental setup and the applied strategy to obtain and optimize the spectroscopy signals. Then we present our data and the used procedure to assign the observed resonances. We derive from it the long-range potential parameters and the molecule formation rates. We first present the results obtained for ^{40}K$_2^*$, then for ^{6}Li^{40}K*. Finally we conclude and give an outlook.

4.2. Experimental results

4.2.1 Experimental setup

A sketch of the optical setup for the PA experiment is depicted in Fig. 4.5. A PA laser beam is directed onto the atoms trapped inside the dual-species ^{6}Li-^{40}K-MOT (described in chapters 2 and 3) and its frequency is scanned. Simultaneously, the steady-state atom number of each species is recorded via the emitted trap fluorescence. The signature of the formation of ^{6}Li^{40}K* molecules is a decrease of both the ^{6}Li and the ^{40}K fluorescence. The PA laser is scanned red detuned with respect to one of the atomic transitions of ^{40}K (see Fig. 4.3) and has no effect on a single-species ^{6}Li-MOT. The ^{6}Li fluorescence signal thus represents a pure heteronuclear PA spectrum, whereas the ^{40}K fluorescence signal represents the sum of a heteronuclear (^{6}Li^{40}K*) and homonuclear (^{40}K$_2^*$) PA spectrum. The frequency of the PA laser is recorded by a wavelength meter (High Finesse, ref. WS-6) with an absolute accuracy of ± 250 MHz. Additionally, a Fabry-Perot interferometer is used to verify the laser's single-mode operation.

The PA light is derived from a homemade diode laser-tapered amplifier system. It has a wavelength of 767 nm and a power of 660 mW at the output of a single-mode polarization-maintaining fiber. The linearly polarized collimated beam has a $1/e^2$-diameter of 2.2 mm and passes four times through the center of the MOT (making use of polarization optics, see Fig. 4.5), with a total peak light intensity of ~ 100 W/cm^2 (taking into account the power losses in the optics). The beam diameter was chosen to match the size of the ^{6}Li-MOT. The beam frequency is scanned by changing the length of the diode laser's external cavity with a piezo actuator. Using the feed-forward technique [220], the laser's mode hop free continuous tuning range extends over ~ 35 GHz. In the experiment the PA laser is scanned over a total range of 325 GHz. To cover this wide range manual tunings of the laser cavity were occasionally required.

4.2.2 Optimization of the photoassociation signal

In this section we describe the various procedures we followed in order to optimize the experimental conditions which lead to the observation of the first PA signal for LiK*. We focused on optimizing the PA signal imprinted on the Li fluorescence as that represents the pure heteronuclear LiK* spectrum allowing for an easier analysis. The first important step was to optimize the stability of the ^{6}Li-MOT fluorescence. We obtained a fluorescence long term stability of $\pm 3\%$, which could be achieved mainly by carefully adjusting the various gains of the feedback electronics used to frequency-stabilize the diode laser used for the ^{6}Li-MOT. The second important step was to slowly scan the frequency of the PA beam (with ~ 15 MHz/s) such that a quasi-steady state for the trap

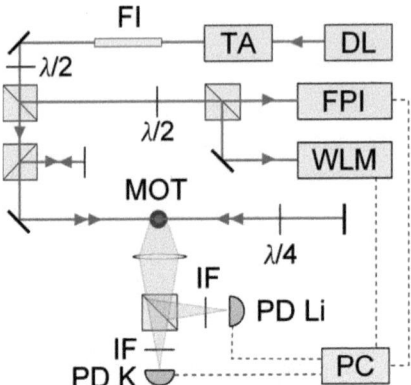

Figure 4.5: Optical setup for the PA spectroscopy experiment. A laser beam originating from a 767 nm diode laser (DL) is amplified by a tapered amplifier (TA), passes through a polarization maintaining single-mode optical fiber (FI) and is directed onto the dual-species MOT. Its frequency is scanned red detuned with respect to one of the atomic transitions of ^{40}K. PA spectra are recorded by continuously measuring the frequency of the DL with a wavelength meter (WLM) and a Fabry-Perot interferometer (FPI) and by continuously recording the atom number in the two MOTs via their emitted fluorescence using two different photo diodes (PD) and narrow band interference filters (IF). In order to achieve high PA laser intensities, the PA beam is retro-reflected two times, such that it passes four times through the MOT.

loss can be obtained (see Fig. 4.16 (a)). The cruicial step was to reduce the ^{6}Li-MOT to a small atom number and volume (by lowering the loading rate by decreasing the ^{6}Li oven temperature) and to place it at the center of the larger ^{40}K-MOT. This could be achieved by independently adjusting the beam alignment of the MOTs while continuously imaging the trapped atoms from two different directions and by monitoring the collision-induced influence of the ^{40}K-MOT on the atom number in the ^{6}Li-MOT. When the influence is maximum (for given MOT-light intensities and frequencies) the overlap is best. Finally, the PA-induced trap loss in the ^{6}Li-MOT needed to be maximized and all other intrinsic losses that compete with it minimized. An understanding of how this is achieved can be obtained from the rate equation for the total number $N_{\text{Li}}^{\text{PA}}$ of trapped ^{6}Li atoms. Assuming that the ^{6}Li-MOT is entirely contained inside the ^{40}K-MOT and that the atom density is uniform (which is a reasonable assumption for our large atom number MOTs), the rate equation is given by [186]

$$\frac{dN_{\text{Li}}^{\text{PA}}}{dt} = L_{\text{Li}} - \left[\gamma + \beta_{\text{LiLi}} n_c^{\text{Li}} + (\beta_{\text{KLi}} + \beta_{\text{PA}}) n_c^{\text{K}}\right] N_{\text{Li}}^{\text{PA}}, \qquad (4.21)$$

4.2. Experimental results

where n_c^{Li}, n_c^K are the constant central trap densities of the respective MOTs, L_{Li} is the ^{6}Li-MOT loading rate, γ the trap loss rate due to background gas collisions. β_{LiLi}, β_{KLi} and β_{PA} represent the trap loss coefficients for atom loss induced by the trapping and the PA light, respectively. We are interested in obtaining a large PA-induced *relative* atom loss $\Delta N_{Li}^{PA}/N_{Li}$, where N_{Li} denotes the ^{6}Li-MOT atom number in the absence of the PA laser and $\Delta N_{Li}^{PA} = N_{Li} - N_{Li}^{PA}$. In steady state (*i.e.* $dN_{Li}^{PA}/dt = 0$), it is

$$\frac{\Delta N_{Li}^{PA}}{N_{Li}} = \frac{\beta_{PA} n_c^K}{\gamma + \beta_{LiLi} n_c^{Li} + (\beta_{KLi} + \beta_{PA}) n_c^K}. \tag{4.22}$$

Thus, the relative PA-induced trap loss is maximized by increasing the PA loss term $\beta_{PA} n_c^K$ and by reducing all other loss terms. The PA loss term $\beta_{PA} n_c^K$ is maximized by using a large MOT magnetic field gradient, which increases $n_c^{K\,3}$, and by using a large light power in the PA beam (by implementing several beam reflections (see Fig. 4.5)), which increases β_{PA}. The PA beam size was optimized using a variable beam expander and found to be optimum for a $1/e^2$-diameter of 2.2 mm which matches the size of the ^{6}Li-MOT.

The loss terms β_{LiLi}, β_{KLi} for light-induced intra- and interspecies collisions are minimized by reducing the light intensity and increasing the detuning of the trapping lasers. For the ^{6}Li-MOT cooling light, however, we found it better not to increase but rather decrease the detuning so as to lower the trap depth of the ^{6}Li-MOT and to decrease the probability for recapture of dissociating excited molecules.

The optimum light detunings and intensities per MOT beam and axial magnetic field gradient were found to be $\Delta\nu_{cool}^{Li} = \Delta\nu_{rep}^{Li} \sim -3\Gamma$, $\Delta\nu_{cool}^{K} = \Delta\nu_{rep}^{K} \sim -4\Gamma$, and $I_{cool}^{Li} \sim 1.5 I_{sat}^{Li}$, $I_{rep}^{Li} \sim 0.5 I_{sat}^{Li}$, $I_{cool}^{K} \sim 10 I_{sat}^{K}$, $I_{rep}^{K} \sim 3 I_{sat}^{K}$ and $\partial_z B = 20$ G/cm, respectively. These parameters result in $N_{Li} \sim 5 \times 10^8$ and $N_K \sim 2.5 \times 10^9$ trapped atoms with central atomic densities of $n_{Li} \sim 7 \times 10^{10}$ cm^{-3} and $n_K \sim 5 \times 10^{10}$ cm^{-3} and temperatures of $T_{Li} \sim 1.2$ mK and $T_K \sim 300\,\mu$K, respectively. For the chosen detunings and intensities, $\sim$30% of the ground-state ^{6}Li atoms are expected to populate the $F = 1/2$- and $\sim$70% the $F = 3/2$-state. For ^{40}K, the relative ground-state populations are expected to be $\sim$20% for the $F = 7/2$-state and $\sim$80% for the $F = 9/2$-state.

When looking for the first heteronuclear PA signal, we did not know at which detunings PA resonances would appear and would have a detectable contrast. We thus strategically chose to scan the PA laser in a region, in which both the density of resonances is expected to be high and the density of the ^{40}K atoms in the trap. At

---
[3]This also increases n_c^{Li} and thus the last three terms in the denominator of Eq. (4.22). However, due to the presence and significance of the fourth term γ in the denominator, the relative PA-induced loss generally increases when the atom densities are increased.

118   Chapter 4. Photoassociation of heteronuclear $^6Li^{40}K$ molecules

very small PA detunings (0-10 GHz) the resonance density is highest, but the density of the ^{40}K atoms is small since the PA laser pushes the atoms out of the MOT. The atom density is highest at intermediate detunings (10-20 GHz), where attractive optical dipole forces [221] due to the intense PA beam outweigh its pushing effect, enhancing the atom density (see Fig. 4.6). Since at these intermediate detunings the PA resonance density is still expected to be high, we searched (and also found) the first heteronuclear PA signal in this region.

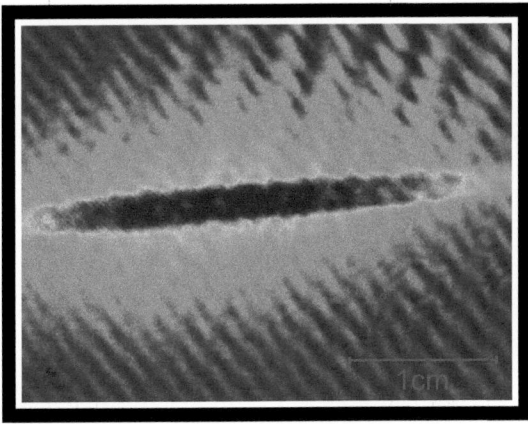

Figure 4.6: Absorption image of the ^{40}K-MOT in presence of the PA beam which is detuned by $\sim$ 15 GHz. The central line represents a spatial region with enhanced density of ^{40}K atoms which results from the attractive optical dipole forces induced by the PA beam. Due to this density enhancement the PA induced heteronuclear losses are expected to be large in this region of PA detunings.

Once the heteronuclear PA signal was detected it could be further optimized following the above-mentioned guidelines, finally leading to a heteronuclear PA-induced trap loss in the ^{6}Li-MOT of up to 35% (see Fig. 4.11 (b)).

4.2.3 Photoassociation spectroscopy of $^{40}K_2^*$ molecules

In a first experiment, we recorded a homonuclear $^{40}K_2^*$ PA spectrum. In this experiment the ^{6}Li-MOT was not present so not to complicate the recorded spectrum by the additional resonances due the formation of $^6Li^{40}K^*$ molecules. Photoassociation of the fermionic potassium isotope ^{40}K has not been demonstrated in the past. For potassium PA has been demonstrated only for the isotope ^{39}K [186, 171]. The difference in mass of the two isotopes leads to different resonance positions in the spectra (see e.g. Eq. (4.14)). However, the molecular potentials and thus also the long-range dispersion

4.2. Experimental results

coefficients are the same for both isotopes, as they result from electrostatic interactions only. This allows us to directly compare our results to the ones of Ref. [186].

Photoassociation spectrum and resonance assignment

Figure 4.7 shows the homonuclear PA spectrum of ^{40}K$_2^*$ near the dissociation limit $4S_{1/2}+4P_{3/2}$ for PA detunings between 0 and $-220\,\text{GHz}$. The graph represents the average of approximately five individually recorded spectra for noise reduction and has been recorded in pieces and stitched together. The spectrum contains more than 130 resonances, belonging to two different vibrational series which have different contrasts. The inset of Fig. 4.7 shows a zoom of the spectrum, where the two series are clearly visible. It can be further seen that the vibrational resonances appear in doublets, which we attribute to a resolved hyperfine structure. A list of all observed resonances and their assignment is presented in Tab. 4.4.

The resonance assignment is based on the following arguments: ten Hund's case (c) potentials dissociate to the asymptote $4S_{1/2}+4P_{3/2}$ (see Fig. 4.8), among which only five potentials support bound states ($0_u^+, 1_g, 0_g^-, 1_u$ and 2_u). The potential 2_u is not optically accessible from the three ground-state potentials $0_g^+, 0_u^-, 1_u$ due to the Hund's case (c) selection rules $\Delta\Omega = 0, \pm 1, g \leftrightarrow u$ mentioned earlier. The potentials 0_g^- and 1_u are purely long-range potentials [222] whose attractive well is entirely located outside the range of the chemical bonding. They have small potential depths of $D(0_g^-) = h \times 194.4\,\text{GHz}$ and $D(1_u) = h \times 16.2\,\text{GHz}$ [223]. Resonances from these potentials are thus not expected to appear for PA detunings $|\Delta_{\text{PA}}| > 195\,\text{GHz}$. The two series observed in our experiment can be excluded to belong to these potentials, as they both contain resonances for PA detunings $|\Delta_{\text{PA}}| > 195\,\text{GHz}$ (see Fig. 4.7). Thus, the two observed series must belong to the potentials 0_u^+ and 1_g. We identify the strong series as belonging to the potential 0_u^+ and the weak series as belonging to 1_g, based on their different level spacings: the potential 1_g has a smaller theoretically predicted C_3 dispersion coefficient and thus its resonances are more widely spaced.

In both vibrational series the resonances appear in doublets of one strong and one weak line, separated by $\sim 1.2\,\text{GHz}$ throughout the spectrum (see inset of Fig. 4.7). The appearance of the doublets originates from the simultaneous population of ^{40}K atoms of both hyperfine ground states $4S_{1/2}(F=7/2)$ and $4S_{1/2}(F=9/2)$ in the MOT. The stronger line of each doublet corresponds to an excitation from the $4S_{1/2}(F=9/2)$ state as it is located at lower PA detunings (see Fig. 2.3). The contrast of the strong line is ~ 5 times larger than of the weak line. Assuming the same oscillator strength for excitations from each of the hyperfine ground states (which is reasonable due to the high PA beam intensity), the difference in contrast indicates that $\sim 80\%$ of the

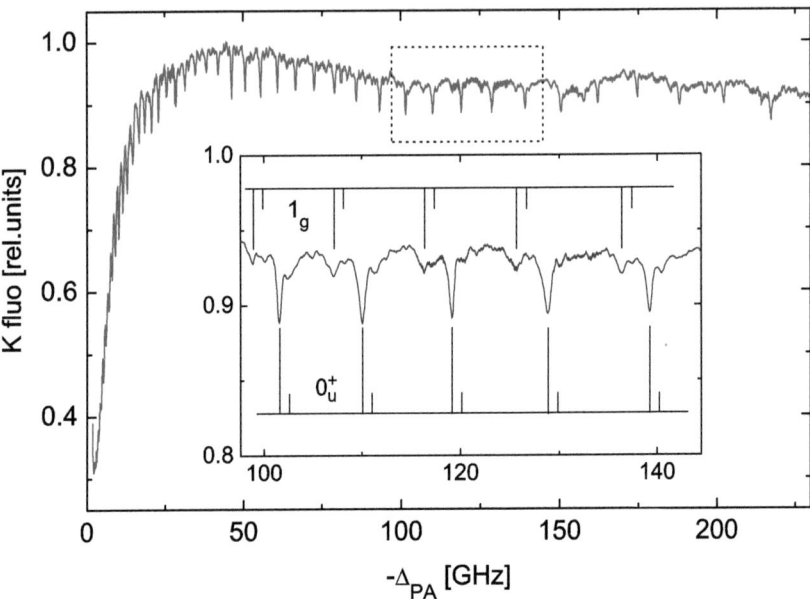

Figure 4.7: Partial homonuclear PA trap loss spectrum of ^{40}K$_2^*$ below the $4S_{1/2}+4P_{3/2}$ asymptote. The PA detuning Δ_{PA} is specified relative to the atomic transition $4S_{1/2}(F = 9/2) \rightarrow 4P_{3/2}(F' = 11/2)$. The spectrum contains two vibrational series belonging to the potentials 0_u^+ and 1_g, whose assignment is given in Tab. 4.4. The inset figure shows a zoom on the spectrum showing five vibrational resonances of each series. Each vibrational resonance is accompanied by a weaker resonance, separated by $\sim 1.2\,\mathrm{GHz}$, due to a significant amount of ^{40}K atoms populating the atomic hyperfine ground state $4S_{1/2}(F = 7/2)$. Rotational splittings are not resolved. For small PA detunings the fluorescence of the ^{40}K-MOT is significantly perturbed by the PA laser, as that pushes the atoms out of the trap.

non-excited trapped ^{40}K atoms were occupying the $4S_{1/2}(F = 9/2)$-state in the MOT during the experiment.

The resonances shown in Fig. 4.7 have a width of $\sim 400\,\mathrm{MHz}$ (FWHM) and they do not show a further substructure. Since the spectrum of Fig. 4.7 was obtained by averaging many recorded spectra, the width of the resonances of the averaged spectrum is affected by the uncertainty of the frequency measurement by the wavelength meter. When no averaging is performed the spectral resolution is better. For example the measured width of the resonance of the 0_u^+ potential at $-50\,\mathrm{GHz}$ detuning when no averaging is performed is of the order of $150\,\mathrm{MHz}$. The main contributions to this width are the excited atomic hyperfine structure ($\sim 80\,\mathrm{MHz}$) and the unresolved rota-

4.2. Experimental results

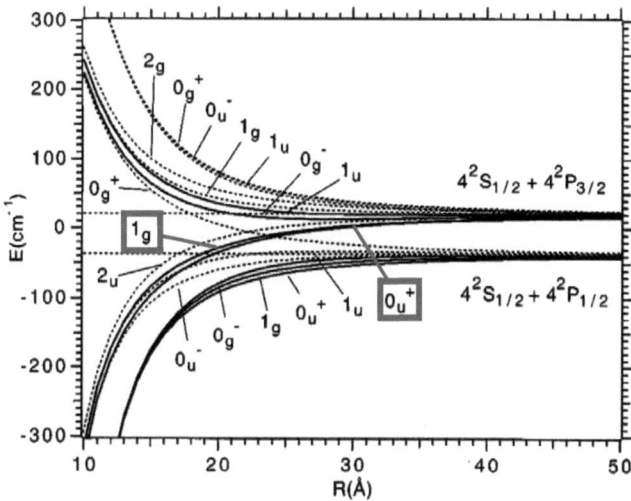

Figure 4.8: Molecular potentials of the K_2 molecule for large interatomic separations R (in units of Å = 10^{-10} m = 1.8897 a_0) dissociating to the $4S_{1/2}+4P_{1/2,3/2}$ asymptotes. The solid curves are those states which support bound states and are accessible by dipole transitions from the ground state. The potentials 0_u^+ and 1_g, dissociating to the $4S_{1/2}+4P_{3/2}$ asymptote, are accessed in our experiment. The potential curves have been taken from Ref. [224].

tional splittings, which we estimate as follows. We have shown in the introduction, that the excited molecules can have a total angular momentum of $J \leq 3$. The rotational energies being given by $E_{\rm rot} = B_v[J(J+1) - \Omega^2]$, the difference between neighboring rotational levels progresses as $2B_v, 4B_v, 6B_v$. The rotational constant B_v for the detuning of -50 GHz can be estimated by its classical value $B_v \sim \hbar^2/(2\mu R_{v+}^2)$ [4], with $R_{v+} \sim 120\, a_0$, leading to $B_v \sim h \times 7$ MHz. Thus, the maximum rotational splitting is of the order of 40 MHz. Another contribution to the width of the resonances is thermal broadening (~ 6 MHz). The resonances of the 0_u^+ potential might additionally be broadened due to predissociation at a short internuclear distance to the lower dissociation limit $4S_{1/2}+4P_{1/2}$ due to spin-orbit mixing with the $A^1\Sigma_u^+$ potential [205, 186]. We did not intend to study the widths of the resonances, as our main goal was the determination of the long-range part of the excited molecular potentials. For this purpose the averaging of the recorded spectra as described above gives the most precise determination of the resonance positions.

---

[4]This estimation holds only for long-range molecules for which $\langle R \rangle \sim R_{v+}$, which is the case for the K_2^* molecules created in our experiment, but not for the shorter ranged LiK* molecules.

Ω	Δv	$-\Delta_{\mathrm{PA}}$ [GHz]	Contr. [%]	$\langle R\rangle$ [a_0]	Ω	Δv	$-\Delta_{\mathrm{PA}}$ [GHz]	Contr. [%]	$\langle R\rangle$ [a_0]
0_u^+	0	2.85	5.0	321	(0_u^+)	37	150.81	4.7	86
	1	3.26	8.3	307		38	162.55	3.9	84
	2	3.85	7.1	291		39	175.00	4.3	82
	3	4.67	8.7	273		40	188.49	3.5	80
	4	5.44	11.3	259		41	202.43	3.6	78
	5	6.25	9.9	247		42	217.65	4.0	76
	6	7.19	9.4	236		43	233.61	3.4	74
	7	8.17	10.1	226		44	250.42	4.2	72
	8	9.23	11.3	217		45	268.41	2.1	71
	9	10.33	12.8	209		46	287.48	2.6	69
	10	11.64	10.8	201		47	308.55	2.5	67
	11	13.02	10.5	194	1_g	0	44.24	1.0	126
	12	14.84	8.6	185		1	48.65	1.3	122
	13	16.86	7.7	178		2	53.18	1.3	119
	14	18.65	6.7	172		3	57.88	1.3	115
	15	20.87	8.4	166		4	63.86	1.4	112
	16	23.04	7.7	160		5	69.81	0.8	108
	17	25.62	4.8	155		6	76.39	1.8	105
	18	28.45	6.4	149		7	83.22	2.0	102
	19	31.40	4.6	144		8	90.96	1.2	99
	20	34.68	3.6	140		9	98.79	1.3	97
	21	38.25	3.9	135		10	107.17	1.4	94
	22	42.25	3.9	131		11	116.33	1.6	91
	23	46.71	8.5	127		12	125.57	1.3	89
	24	50.97	6.7	123		13	136.39	1.3	87
	25	55.81	7.5	119		14	147.51	1.3	85
	26	61.14	6.9	116		15	158.94	1.2	82
	27	66.87	5.2	112		16	171.81	1.4	80
	28	72.76	4.7	109		17	185.36	1.6	78
	29	79.27	5.0	106		18	199.40	1.4	76
	30	86.26	5.6	103		19	214.32	1.3	75
	31	93.70	5.6	100		20	230.67	1.3	73
	32	101.78	5.4	98					
	33	110.34	4.8	95					
	34	119.43	5.2	93			Accuracy		
	35	129.09	5.2	90			±0.25		±1.0
	36	139.50	4.5	88					

Table 4.4: PA resonances of ^{40}K$_2^*$ observed below the $4S_{1/2}+4P_{3/2}$ asymptote. Only the resonances corresponding to excitations from the hyperfine ground state $4S_{1/2}(F=9/2)$ are listed. $\Delta v = v-v_{\min}$ denotes the relative vibrational quantum number. The average radii of the molecules are estimated by $\langle R\rangle \sim R_{v+} = [-C_3/(h\Delta_{\mathrm{PA}})]^{1/3}$.

4.2. Experimental results   123

Determination of the long-range potential

Having assigned all observed resonances, the C_3 coefficients for the excited molecular potentials can be derived. Therefore we employ the LRB formula (Eq. (4.14)) introduced in Sec. 4.1.8, where we have shown that its assumptions are well satisfied for the case of K_2^* for the investigated range of PA detunings (except for very small PA detunings). We can thus expect to obtain high-precision estimates for the C_3 coefficients. Fig. 4.9 shows the plots of the 1/6-rd power of the measured binding energies $D - E_v = -\hbar\Delta_{\mathrm{PA}}$ as a function of the relative vibrational quantum number $\Delta v = v - v_{\min}$ for the two observed series, where v is the absolute value of the vibrational quantum number counted from the dissociation limit and $v_{\min}$ its smallest observed value (v cannot be determined precisely due to the high density of resonances close to the atomic resonance). The plots are predicted by Eq. (4.14) to follow straight lines whose slopes yield: $C_3(0_u^+) = 14.20 \pm 0.02$ a.u. and $C_3(1_g) = 13.37 \pm 0.05$ a.u. Only vibrational levels with binding energies larger than $h \times 40$ GHz have been considered for the fit in order to fulfill the assumptions of the LRB formula (see Sec. 4.1.8). The determined values compare very well with the experimental values $C_3^{\mathrm{W}}(0_u^+) = 14.14 \pm 0.05$ a.u. and $C_3^{\mathrm{W}}(1_g) = 13.54 \pm 0.10$ a.u. found by Wang et al. [186] for $^{39}\mathrm{K}_2^*$. They also compare well with the theoretical values 14.44 and 13.41 derived from the dipole matrix element $d = \langle 4S|\mu|4P\rangle$ for the atomic transition $4S \to 4P$ calculated by Marinescu and Dalgarno [204] using the relations (in atomic units) $C_3(0_u^+) = 5d^2/3$ and $C_3(1_g) = (\sqrt{7}+2)d^2/3$, valid for the observed Hund's case (c) states [205]. The uncertainties represent the statistical uncertainties for the fits, which are found to be of the same order as the uncertainty arising from the determination of the PA detuning by the wavelength meter.

Determination of the lifetime of the excited atomic state $4P$

The dispersion coefficients being linked to the dipole matrix element d, they allow the determination of the lifetime τ of the atomic state $4P$ of $^{40}\mathrm{K}$, which is given by the relation [174, 225]

$$\tau = \frac{3\pi\hbar\varepsilon_0 c^3}{d^2\omega^3}, \qquad (4.23)$$

where $\omega = 2.4568 \times 10^{15}\,\mathrm{s}^{-1}$ is the angular frequency of the atomic transition [226] and ε_0 the electric constant. The average value of d^2 obtained in our experiment is $d^2 = 8.577 \pm 0.02$ a.u., yielding a lifetime of $\tau = 25.94 \pm 0.07$ ns (which corresponds to a linewidth $\Gamma/(2\pi) = 6.136 \pm 0.017\,\mathrm{MHz}$). This value is slightly smaller than the

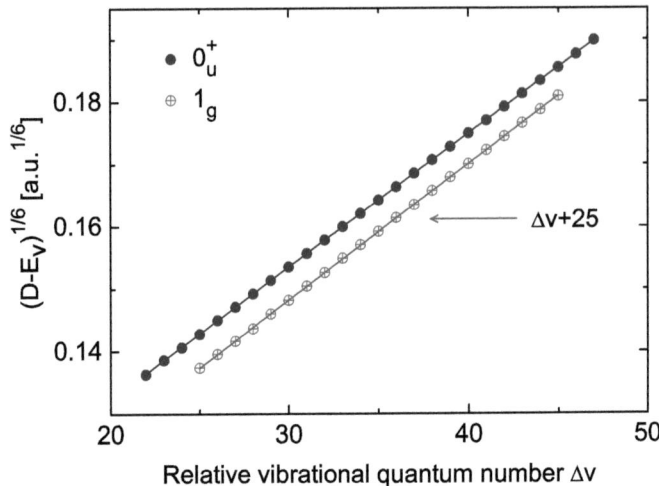

Figure 4.9: Plot of the 1/6-th power of the measured binding energies $D - E_v = -h\Delta_{\text{PA}}$ (symbols) as a function of the relative vibrational quantum number Δv (see Tab. 4.4) for the two observed vibrational series 0_u^+ and 1_g of $^{40}\text{K}_2^*$ dissociating to the $4S_{1/2}+4P_{3/2}$ asymptote. Δv has been shifted by 25 for the data of the 1_g series for demonstration purposes. The slopes of the linear fits (solid lines) yield the dispersion coefficients C_3 according to the LRB formula (Eq. (4.14)). Only the resonances at large detunings ($-\Delta_{\text{PA}} > 40\,\text{GHz}$) are used for the fit, since the LRB formula is not valid for the most weakly bound molecular states.

currently most precise value of 26.34 ± 0.05 ns (which corresponds to a linewidth $\Gamma/(2\pi) = 6.042 \pm 0.011\,\text{MHz}$), which has been determined by PA spectroscopy of $^{39}\text{K}_2^*$ [225]. We attribute this discrepancy to a systematic error (not considered by the uncertainty given above) in our result, which arises from the application of the LRB formula. In the experiment of Ref. [225], a smaller uncertainty for the dipole matrix element could be obtained by deriving it from the dispersion coefficient of a purely long-range potential. Such potentials allow a more precise determination of the corresponding dispersion coefficient, since they are nearly not affected by exchange interactions [225]. Furthermore, the PA detuning in this experiment was measured with an eight-times higher precision of 30 MHz, using two thermally stabilized Fabry-Perot etalons. We did not invest in an ultra-precise determination of the PA detuning, which could be realized using femtosecond laser combs. In our experiment we could not unambiguously identify excitations to purely long-range molecular potentials. The observation of such excitations is known to be difficult, because on the one hand, the small depth of these potentials very likely leads to dissociation to low-energy and thus

4.2. Experimental results

recapturable atoms, not detectable by the trap loss technique. On the other hand the potentials only contain weakly bound states, which tend to be hidden in the PA spectrum as they appear for small PA detunings, where the state density of other potentials with stronger signals is also high.

Conclusion

In this section we have presented our experimental results on photoassociation of homonuclear ^{40}K$_2^*$ molecules. We have observed and identified two vibrational series and we have determined the leading order dispersion coefficients of the corresponding excited molecular potentials making use of the LRB formula. The obtained values are in very good agreement with those found by Wang et al. [186] for ^{39}K$_2^*$ and with theoretical predictions. Using our values we could derive the lifetime of the atomic state $4P$ of ^{40}K with a precision of 0.3%, close to the currently most precise value [225].

4.2.4 Photoassociation spectroscopy of ^{6}Li^{40}K* molecules

In the following, we present our results on photoassociation of heteronuclear ^{6}Li^{40}K* molecules. We have recorded two PA spectra: one in which the PA laser is detuned close to the $2S_{1/2}+4P_{3/2}$ asymptote and one in which it is detuned close to the $2S_{1/2}+4P_{1/2}$ asymptote (see Fig. 4.3). The first spectrum was recorded for PA detunings between 0 and -325 GHz, the second for PA detunings between 0 and -60 GHz. For higher detunings, no further resonances could be observed with a contrast large enough to allow for a clear distinction from noise.

Photoassociation spectra and resonance assignment

The recorded heteronuclear PA spectrum close to the $2S_{1/2}+4P_{3/2}$ asymptote (as appearing on the Li fluorescence signal) is shown in Fig. 4.10 (the spectrum recorded below the $2S_{1/2}+4P_{1/2}$ asymptote is shown in Fig. 4.13 and will be discussed later). The graph is an average of ~ 6 recorded spectra for noise reduction and has been recorded in pieces and stitched together. Two weeks of continuous data acquisition were required to obtain the spectrum. The spectrum contains 60 resonances whose contrasts decrease and whose mutual separations increase with increasing detuning. The maximum contrast amounts to $\sim 35\%$ and is obtained for a detuning of $\Delta_{\text{PA}} = -14.4$ GHz. The observed resonance widths (FWHM) vary between 80 and 300 MHz, primarily due to unresolved molecular hyperfine structure.

We have also recorded the heteronuclear+homonuclear PA spectrum appearing on the K fluorescence signal, which contains all the resonances of Fig. 4.10 as well. A comparison between the two spectra is shown for a small part in Fig. 4.11 (a). This figure shows comparable contrasts for the heteronuclear ^{6}Li^{40}K* and homonuclear ^{40}K$_2^*$ PA signals. The homonuclear ^{40}K$_2^*$ resonances visible on the K fluorescence (lower trace, right axis) are the previously identified resonances belonging to the series 0_u^+ (see Tab. 4.4).

In the heteronuclear spectrum of Fig. 4.10 we identify five vibrational series (labeled with numbers), corresponding to the five molecular potentials dissociating to the $2S_{1/2}+4P_{3/2}$ asymptote (see Fig. 4.3 b)). Each series contains five resonances, which appear in doublets or triplets due to resolved rotational structure (see Tab. 4.3). This structure is shown more clearly in the high-resolution spectrum of Fig. 4.11 (a) (upper trace, left axis) for the $\Omega = 1^{\mathrm{up}}, v = 3$ vibrational state. Some of the observed rovibrational resonances have a substructure resulting from hyperfine interactions, which is shown in the high-resolution spectrum of Fig. 4.11 (b) for the resonance $\Omega = 1^{\mathrm{up}}, v = 2, J = 1$.

Before attempting to assign the observed resonances, it should be verified that all resonances are due to the formation of singly excited heteronuclear molecules. The heteronuclear character of the resonances has already been confirmed by their appearance on both the Li and the K fluorescence signal [5]. Since in the ^{6}Li-MOT, both ground-state and excited ^{6}Li atoms are present, it is, however, not *a priori* excluded that doubly excited ^{6}Li^{*40}K* molecules are formed [197]. Even though the fraction of excited atoms in the ^{6}Li-MOT is small (~ 0.1), the rate of formation of doubly excited ^{6}Li^{*40}K* molecules could be large due to the long-range character of the corresponding molecular potential, which scales with the internuclear separation as $1/R^5$ [167]. In order to show that the observed resonances are only due to the formation of singly excited molecules, we recorded a second PA spectrum for a different experimental setting, in which both the ^{6}Li-MOT trapping beams (cooling and repumping) and the PA laser were periodically chopped out of phase with the same frequency, as shown in Fig. 4.12 (a). The chopping period was chosen large compared to the lifetime of the excited ^{6}Li atoms but not too large to prevent the atoms from leaving the MOT. In this setting no excited ^{6}Li atoms are present when the PA laser is on. Figure 4.12 (b) shows the obtained "pulsed PA" spectrum together with the spectrum of Fig. 4.10 for comparison. One can see that most of the resonances appear in both spectra. Some resonances are, however, not present in the "pulsed PA" spectrum. This might, however, simply be due to the reduced contrast which results from the lower average PA

---

[5] These resonances can be assumed to be due to the formation of diatomic molecules, because the probability for the creation of molecules containing more than two atoms is negligible for the small atomic densities in the MOTs.

4.2. Experimental results

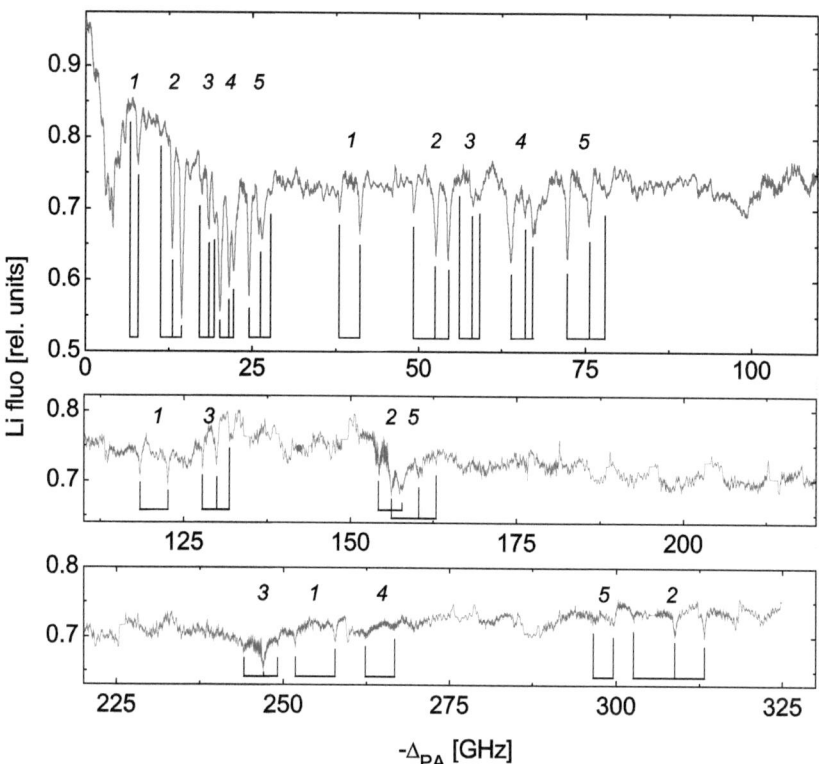

Figure 4.10: Heteronuclear PA trap loss spectrum of ^{6}Li^{40}K* below the $2S_{1/2}+4P_{3/2}$ asymptote. The PA detuning Δ_{PA} is specified relative to the $4S_{1/2}(F = 9/2) \rightarrow 4P_{3/2}(F' = 11/2)$ transition of ^{40}K. The spectrum contains five vibrational series (labeled $N = 1,...,5$) with resolved rotational structure, whose assignment is given in Tab. 4.5. For small PA detunings the average fluorescence of the ^{6}Li-MOT is higher, since it is less influenced by the ^{40}K-MOT, which has a reduced atom number due to the perturbation induced by the near resonant PA laser.

power, that also hindered us to record the pulsed PA spectrum for higher detunings. Nonetheless, the experiment allows us to conclude that most of the observed resonances in the investigated range of PA detunings can be attributed to the formation of singly excited ^{6}Li^{40}K* molecules. Since all of these resonances show vibrational progressions which correspond to the theoretical predictions for singly excited ^{6}Li^{40}K* molecules, as we will see in the next paragraph, it is very likely that all resonances are due to the creation of singly excited ^{6}Li^{40}K* molecules.

Having verified the character of the observed resonances, we now turn to their

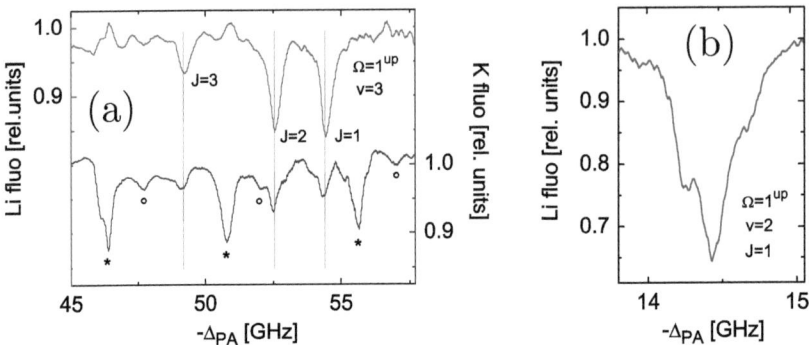

Figure 4.11: (a) Zoom on the heteronuclear (upper trace, left axis) and heteronuclear+homonuclear (lower trace, right axis) PA spectrum below the $2S_{1/2}+4P_{3/2}$ asymptote showing the rotational structure of the $\Omega = 1^{\mathrm{up}}, v = 3$ vibrational state of $^6\mathrm{Li}^{40}\mathrm{K}^*$ and three vibrational 0_u^+ states of $^{40}\mathrm{K}_2^*$, which show a resolved hyperfine (∗ and ○) but no rotational structure. (b) Zoom on the $\Omega = 1^{\mathrm{up}}, v = 2, J = 1$ resonance of $^6\mathrm{Li}^{40}\mathrm{K}^*$, showing a nearly resolved hyperfine structure.

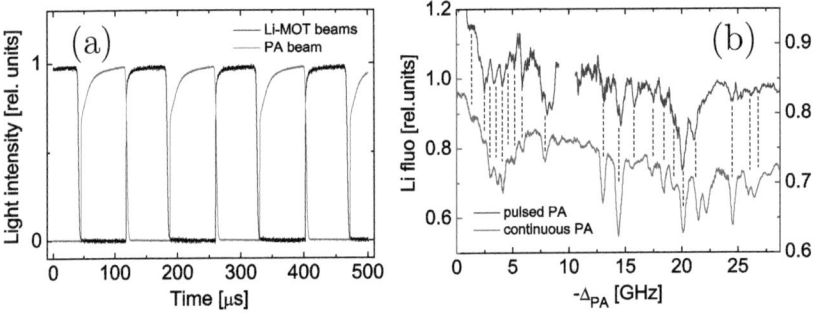

Figure 4.12: (a) Imposed temporal behavior of the light intensity of the $^6\mathrm{Li}$-MOT and the PA beams during data acquisition for the "pulsed PA" spectrum shown in Fig. (b) together with the habitual "continuous PA" spectrum for comparison. The $^6\mathrm{Li}$-MOT and the PA beams being pulsed out of phase, PA can occur only when the atoms in the $^6\mathrm{Li}$-MOT are in their electronic ground state, preventing the formation of doubly excited heteronuclear $^6\mathrm{Li}^{*40}\mathrm{K}^*$ molecules. Figure (b) shows that nearly all resonances of the continuous PA spectrum (lower trace, left axis) appear also in the pulsed PA spectrum (upper trace, right axis), indicating that the resonances are due to the formation of singly excited molecules.

assignment. The heteronuclear spectrum of Fig. 4.10 looks much more complex than the one obtained for K_2^* (see Fig. 4.7). The assignment of the resonances is quite involved. The K_2^*-spectrum could be understood only on the basis of the LRB formula.

4.2. Experimental results

This approach is insufficient for the case of LiK* for the following reasons. First, the LRB formula is much less precise due to the small internuclear distances at which the LiK* molecules are formed (see Sec. 4.1.8). Besides, significant discrepancies exist between the different theoretical predictions for the C_6 coefficients (see Tab. 4.2), which yield an incertitude about the values to be used. Second, due to the small mass of LiK*, the rotational splittings are very large and of the order of the observed vibrational splittings and thus complicate the identification of vibrational series. This identification is further complicated by the resulting significant perturbations between rotational levels of different potentials [227] which lead to irregular rotational structures. Finally, as heteronuclear molecules have less restrictive selection rules, more vibrational series are observed which increases the complexity of the spectrum.

In Fig. 4.10 and Tab. 4.5 I present an assignment of the observed resonances. The assignment was made possible by the combination of several assignment rules: first, the rotational progression law given by Eq. (4.9) combined with our theoretical calculations of the rotational constants B_v. With the help of this law rotational progressions (see Tab. 4.3) can be identified and some of the quantum numbers J and Ω can be assigned based on the observed line spacings. The identification of the lines belonging to the $\Omega = 2$ vibrational series (series 1 in Fig. 4.10) is particularly easy, because for this series only two rotational lines per vibrational level are expected, as opposed to three for all other series (see Tab. 4.3). Second, the LRB law combined with the theoretical predictions of the C_6 coefficients. It predicts that potentials with the same C_6 coefficient have the same vibrational progressions (*i.e.*, the difference between the 1/3-rd power of the binding energies of any neighboring vibrational states is the same for these potentials), and that the vibrational levels of potentials with smaller C_6-values progress faster with the PA detuning. With the help of the LRB law vibrational progressions can be identified and some of the quantum numbers v and Ω can be assigned based on the observed line spacing. Third, the hyperfine structure law $E_{\text{hfs}} \propto \Omega/[J(J+1)]$ for $\Omega = 1$ and $E_{\text{hfs}} \approx 0$ for $\Omega = 0$ [228, 219]. It predicts small widths for resonances with $\Omega = 0$ and particularly large widths for those with $\Omega = 1, J = 1$, facilitating their identification. Fourth, the expected similar contrast pattern of rotational lines belonging to the same vibrational series, which facilitates to identify vibrational progressions.

The detailed procedure of the line assignment has been carried out as follows. First, we searched the spectrum for isolated multiplets, which would allow the determination of the quantum number Ω from its rotational spacing. One such multiplet is found at a detuning of $-41.1\,\text{GHz}$. It consists of only two lines, indicating that it belongs to the dyad potential $\Omega = 2$. This indication is further confirmed by the spacing of the lines which corresponds to $\sim 6B_v$, where B_v is our theoretical prediction for the rotational

N	Ω	v	J	$-\Delta_{PA}$ [GHz]	Contr. [%]	$\langle R \rangle$ [a_0]	N	Ω	v	J	$-\Delta_{PA}$ [GHz]	Contr. [%]	$\langle R \rangle$ [a_0]
1	2	1	2	0.37	1.1	–	4	0^-	2	2	3.03	21.4	44
		2	3	6.60	2.3	40				1	3.66	22.7	
			2	7.88	9.4				3	2	20.11	25.0	32
		3	3	38.00	5.6	26				1	21.48	20.4	
			2	41.08	10.6					0	22.16	17.4	
		4	3	118.49	6.3	22			4	2	63.99	14.7	25
			2	122.67	6.3					1	66.06	5.0	
		5	3	251.80	3.7	19				0	67.13	7.6	
			2	257.81	4.3				6	2	262.37	1.9	22
2	1^{up}	1	2	0.37	1.1	38				0	266.75	1.9	
			1	1.34	6.1		5	1^{down}	2	3	4.12	23.9	42
		2	3	11.27	2.4	33				2	4.85	14.9	
			2	13.01	18.5					1	5.90	4.6	
			1	14.40	35.0				3	3	24.52	19.8	32
		3	3	49.20	6.5	25				2	26.15	9.8	
			2	52.47	14.1					1	27.72	3.0	
			1	54.53	14.8				4	3	72.26	15.4	25
		4	2	154.35	5.4	21				2	75.55	10.0	
			1	157.40	4.7					1	77.87	5.0	
		5	3	302.50	2.0	18			5	3	156.12	6.1	22
			2	308.75	3.9					2	160.20	1.4	
			1	313.26	4.6					1	162.91	1.9	
3	0^+	2	2	1.34	6.1	35			6	2	296.51	1.8	21
			1	2.35	11.0					1	299.57	1.8	
			0	3.03	21.4		6	0^+	2	2	8.66	3.9	34
		3	2	17.23	6.2	32				1	9.74	5.8	
			1	18.51	10.6					0	10.37	0.8	
			0	19.25	8.4				3	2	41.56	1.6	27
		4	2	56.30	2.0	26				1	43.40	5.1	
			1	58.17	6.6					0	44.31	2.0	
			0	59.21	5.6		7	1	2	2	20.42	3.6	30
		5	2	127.76	4.0	25				1	21.93	2.0	
			1	129.91	7.8								
			0	131.80	5.0								
		6	2	244.13	2.0	20					Accuracy		
			1	247.07	5.2						±0.25	±1.0	±1
			0	249.01	2.2								

Table 4.5: PA resonances of ^{6}Li^{40}K* observed below the $2S_{1/2}+4P_{1/2,3/2}$ asymptotes and their contrasts. N denotes the number of the vibrational series given in Figs. 4.10 and 4.13, v the vibrational quantum number counted from dissociation. The average radii of the molecules are determined by $\langle R \rangle = \hbar/\sqrt{2\mu B_v}$ using the measured B_v.

4.2. Experimental results

constant for the $\Omega = 2$ potential. Furthermore, the doublet has a partner at -7.9 GHz, which is separated from it by a value which agrees with the one obtained from the LRB formula when the C_6 coefficient for the $\Omega = 2$ potential is used. Furthermore the partner is also a doublet and has a rotational spacing according to the one expected for the $\Omega = 2$ potential. In addition, the LRB formula clearly excludes the doublet at -41.1 GHz to belong to one of the upper triad potentials $0^-, 0^+, 1^{\text{down}}$, since, using their C_6 coefficient, the LRB formula would predict the neighboring vibrational state to appear at a detuning of -10.8 GHz, whereas no resonance can be found in close proximity of this value in the spectrum [6]. We thus conclude that the doublets at -7.9 GHz and -41.1 GHz belong to the dyad potential $\Omega = 2$. An extrapolation based on the LRB formula roughly yields the positions of the remaining resonances of this series (series 1 in Fig. 4.10), which are then distinguished from resonances found in the neighborhood by the knowledge of the rotational splittings. Due to the low density of vibrational states, the LRB formula allows the assignment of the (absolute) quantum number v [7].

Another isolated multiplet is found at a detuning of -14.4 GHz. It consists of three lines whose spacing corresponds to $\sim 6B_v$ and $\sim 4B_v$, where B_v is our theoretical prediction for the dyad potential $\Omega = 1^{\text{up}}$ potential. The triplet has a partner at -54.5 GHz, which is separated from it by a value which agrees with the one obtained from the LRB formula and the C_6 coefficient for the $\Omega = 1^{\text{up}}$ potential. The suggested assignment to the $\Omega = 1^{\text{up}}$ potential is further confirmed by the large width of the rovibrational line at -14.4 GHz (see Fig. 4.11 (b)), which indicates that it is a $\Omega = 1, J = 1$ resonance. An extrapolation with the LRB formula yields the other lines of this series (series 2 in Fig. 4.10), whereas a consideration of the rotational splittings and the relative contrasts of the rotational lines is required to unambiguously distinguish the series from lines of other series in close proximity.

From the already identified series we know that the remaining lines belong to the upper triad potentials $0^-, 0^+, 1^{\text{down}}$, which have all the same C_6 coefficients (see Tab. 4.2). The spectrum in Fig. 4.10 shows three distinct triplets in the region between -17 and -27 GHz. Those have each one partner in the region between -56 and -78 GHz, which is separated by the value expected from the LRB formula. This confirms that these three triplets and their partners belong to the potentials $0^-, 0^+, 1^{\text{down}}$. We number the obtained series by 3, 4 and 5 and we obtain the remaining lines by using the LRB

---

[6]It might be that this resonance is simply suppressed by a node in the initial scattering wave function, which, however, is unlikely since we observe 15 resonances between -6 GHz and -28 GHz (see Tab. 4.5), whereas only one vibrational state per potential is expected in this range, suggesting that all five vibrational series are observed in this range and that no line suppression occurs.

[7]This has not been possible for $^{40}\text{K}_2^*$, see Tab. 4.4.

formula. The assignment of the lines to the series is further confirmed by the constant relative line contrasts of the rotational lines within each series: for series 5, the rotational line with $J = 3$ is the strongest throughout the spectrum and the line with $J = 1$ is the weakest (see Tab. 4.5). For series 3, the rotational line with $J = 1$ is the strongest throughout the spectrum, the lines with $J = 2$ and $J = 0$ being less strong. For series 4, the rotational line with $J = 1$ is the weakest throughout the spectrum.

Figure 4.10 shows that series 1 and 2 of the dyad potentials progress faster with the PA detuning than series 3, 4 and 5 of the upper triad potentials, which is a direct consequence of their larger C_6 coefficients (note how series 1 and 2 "overtake" series 3, 4 and 5).

We now assign the quantum numbers Ω and the partity σ to the series 3, 4 and 5. Series 5 is readily determined as belonging to the potential 1^{down} for the following reasons. First, it has larger rotational splittings than series 3 and 4, and second the lines clearly have a larger width indicating the appearance of a hyperfine structure, which is not expected for the potentials $0^-, 0^+$. The remaining series 3 and 4 thus both belong to potentials with $\Omega = 0$. The determination of their parity is, however, more delicate, since both series have identical C_6, B_v and Ω. It requires an analysis of the relative strength of the rotational lines. Due to the selection rules, the parity of the total wave function of the system, i.e. the product of σ and $(-)^l$ for the rotational part (l being the partial wave), changes sign during the transition. Further, the electronic parity is conserved, namely only $X^1\Sigma^+(0^+) \to 0^+$ and $a^3\Sigma^+(0^-) \to 0^-$ are allowed for parallel transitions. In our experiment, s-wave collisions dominate, such that the total parity is $+$ ($-$) for the former (the latter) initial state. The parallel transition $X^1\Sigma^+(0^+, l = 0) \to (0^+, l = 1)$ is thus allowed promoting then the $J = 1$ line, while the parallel transition $a^3\Sigma^+(0^-, l = 0) \to (0^-, l = 1)$ is forbidden. Under the same approximation, the perpendicular transition $a^3\Sigma^+(1, l = 0) \to (0^-, l = 0)$ is allowed and promotes the $J = 0$ line in the spectrum. Therefore we assign the $\Omega = 0$ series with pronounced (reduced) $J = 1$ line to the excited 0^+ (0^-) state.

In the previous paragraphs we have analyzed the heteronuclear PA spectrum which has been recorded below the $2S_{1/2}+4P_{3/2}$ asymptote. We now turn to the spectrum near the dissociation limit $2S_{1/2}+4P_{1/2}$, which is shown in Fig. 4.13 for PA detunings from 0 to $-60\,\text{GHz}$. The spectrum contains 8 resonances, which have small contrasts of less than 6%. Averaging of ~ 20 recorded spectra was required to make the resonances visible well above noise level. We identify two vibrational series (labeled with numbers) corresponding to two of the three molecular potentials dissociating to the $2S_{1/2}+4P_{1/2}$ asymptote (see Fig. 4.3). Their assignment is presented in Fig. 4.13 and Tab. 4.5 and based on the above-mentioned rules: series 6 (see Fig. 4.13) consists of two triplets with

4.2. Experimental results

vibrational and rotational spacings corresponding to those expected for the potentials with $\Omega = 0$ from the LRB and rotational progression laws. We identify the series to belong to the specific potential $\Omega = 0^+$, due to the relative strength of the rotational lines, which is largest for the $J = 1$ line. Series 7 only contains one doublet whose rotational spacing indicates that it belongs to the potential $\Omega = 1$. Furthermore, its $J = 1$ resonances is significantly broader than all other resonances, indicating the appearance of a hyperfine structure, which confirms the assignment of this line to the potential $\Omega = 1$.

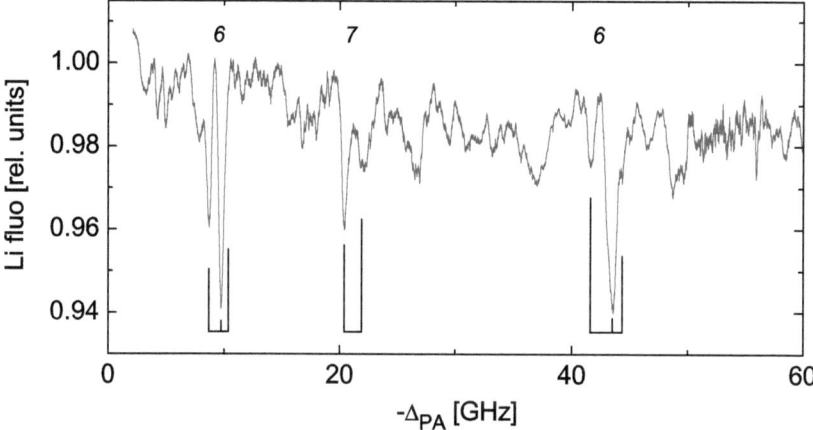

Figure 4.13: Heteronuclear PA trap loss spectrum of ^{6}Li^{40}K* below the $2S_{1/2}+4P_{1/2}$ asymptote. The PA detuning Δ_{PA} is specified relative to the $4S_{1/2}(F = 9/2) \to 4P_{1/2}(F' = 9/2)$ transition of ^{40}K. The spectrum contains two vibrational series (labeled $N = 6, 7$) with resolved rotational structure, whose assignment is given in Tab. 4.5.

Derivation of the long-range potential and the rotational constants

Having assigned all observed resonances, the parameters of the different molecular potentials can be derived. We infer the C_6 coefficients from the measured vibrational binding energies using the LRB formula (Eq. (4.14)). Due to the non-negligible rotational splittings, it is required to subtract the rotational energy from the measured binding energies, *i.e.* to take $D - E_v = -(h\Delta_{\text{PA}} - E_{\text{rot}})$. Figure 4.14 shows the plots of the 1/3-rd power of the binding energies as a function of the vibrational quantum number for the five vibrational series dissociating to the $2S_{1/2}+4P_{3/2}$ asymptote. The plots are predicted by Eq. (4.14) to follow straight lines whose slopes yield:

$C_6 = 9170 \pm 940$ a.u. and $C_6 = 9240 \pm 960$ a.u. for the dyad potentials $\Omega = 2, 1^{up}$, $C_6 = 25220 \pm 600$ a.u., $C_6 = 25454 \pm 720$ a.u. and $C_6 = 24310 \pm 1710$ a.u. for the upper triad potentials $\Omega = 1^{down}, 0^+, 0^-$ and $C_6 = 12860 \pm 660$ a.u. for the lower triad potential $\Omega = 0^+$ (not shown in Fig. 4.14), respectively, where the uncertainties represent statistical uncertainties for the fits. These values are in very good agreement with the respective theoretical values $C_6 = 9800$ a.u., $C_6 = 25500$ a.u. and $C_6 = 13830$ a.u. predicted by Bussery et al. [166] (see Tab. 4.2). The agreement with the values $C_6 = 9520$ a.u., $C_6 = 22000$ a.u. and $C_6 = 15420$ a.u. predicted by Movre et al. [203] and the similar ones predicted by Marinescu et al. [167] is not as good. The three theoretical predictions differ in their treatment of the interaction between the two asymptotes $2S+4P$ and $2P+4S$, which is taken into account in Ref. [166] only. The agreement of our data with this reference hints that this interaction is significant.

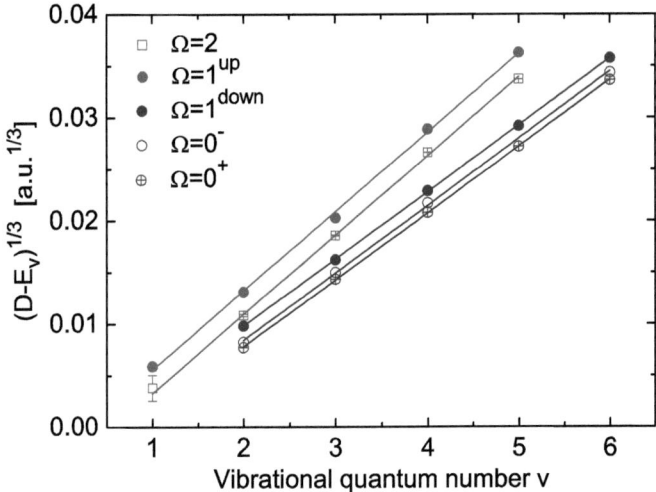

Figure 4.14: Plot of the 1/3-rd power of the measured binding energies $D - E_v = -(\hbar \Delta_{PA} - E_{rot})$ (symbols) as a function of the vibrational quantum number counted from the dissociation limit for the five vibrational series dissociating to the $2S_{1/2}+4P_{3/2}$ asymptote. The slopes of the linear fits (solid lines) yield the dispersion coefficients C_6 according to the LRB formula (Eq. (4.14)). The LRB formula indicates that the least bound vibrational states ($v = 1$) are observed for the potentials $\Omega = 2, 1^{up}$. The identically appearing slopes of the three triad potentials and the two dyad potentials demonstrate the equality of the respective C_6 coefficients.

The uncertainty of the experimentally derived C_6 coefficients results from the following effects. First, the heteronuclear nature of the LiK molecule and its small C_6

4.2. Experimental results

coefficients lead to molecule formation at small internuclear separations (see values of $\langle R \rangle$ in Tabs. 4.5 and 4.4) at which the exchange interaction and higher-order terms in the long-range multipole expansion of the molecular potential (see Eq. (4.5)) become important. As the LRB law neglects these contributions systematic errors arise. Second, the small reduced mass of the LiK molecule leads to a low density of vibrational states [8] and thus to a small number of states with long-range character, causing statistical errors.

The measured rotational splittings allow us to infer the rotational constants and to confirm the assignments above. They are shown in Fig. 4.15 for the five vibrational series below the $2S_{1/2}+4P_{3/2}$ asymptote, together with their theoretical predictions, which we have derived from the potential curves of Fig. 4.3. [9]. The agreement between the measured and theoretically predicted values is reasonable. The error bars account for the imprecision of the wavelength determination and of the resonance positions due to the unresolved hyperfine structure. Deviations from the theoretical predictions may be due to perturbations by the $3^1\Sigma^+$ state [215].

In summary, our experimentally obtained values for the C_6 coefficients and the rotational constants are in very good agreement with the theoretical predictions. This agreement confirms our resonance assignments.

Determination of the molecule formation rate

In the following we determine the formation rate for excited $^6\text{Li}^{40}\text{K}^*$ molecules and we compare it to the one obtained for $^{40}\text{K}_2^*$ in our experiment and to our own theoretical estimates. Besides we study the dependence of the $^6\text{Li}^{40}\text{K}^*$ molecule formation rate on the intensity of the PA light. Finally, we theoretically predict the probabilities for the created excited $^6\text{Li}^{40}\text{K}^*$ molecules to decay into the different vibrational levels of the singlet electronic ground-state potential $X^1\Sigma^+$.

An estimate for the molecule formation rate can be obtained by two different experimental procedures. In the first, the rate is deduced from the steady state atom numbers in the ^{6}Li-MOT in presence and in absence of the PA beam using Eq. (4.21). The large asymmetry in the populations of the two MOTs allows one to consider K a constant background gas, which does not change its density when PA occurs. For the resonance shown in Fig. 4.11 (b) we obtain (at the maximum

---

[8] For example, in the investigated range of PA detunings, 5 vibrational levels per potential were observed for $^6\text{Li}^{40}\text{K}^*$, whereas 48 were observed for $^{40}\text{K}_2^*$.

[9] Even though theoretically calculated potential curves cannot be invoked to predict the binding energies of weakly bound molecular levels, they are well adapted to the prediction of rotational constants, as those depend on the expectation value of the molecule's radius, which does not critically depend on the precise details of the potentials.

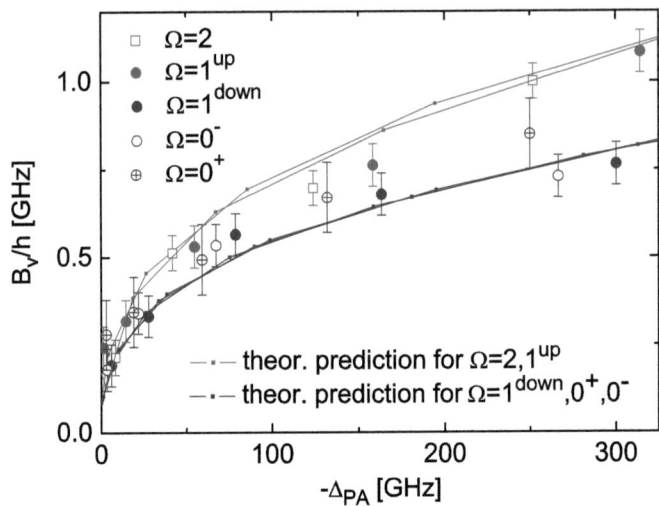

Figure 4.15: Measured rotational constants (symbols) for the observed excited molecular states below the $2S_{1/2}+4P_{3/2}$ asymptote and their theoretical predictions for computed vibrational levels (dots, the lines serve to guide the eye), derived from the potential curves of Fig. 4.3.

available PA laser intensity of $100\,\mathrm{W/cm^2}$) a lower bound for the molecule formation rate of $\beta_{\mathrm{PA}} n_c^K N_{\mathrm{Li}}^{\mathrm{PA}} \sim 3.5 \times 10^7$ molecules/s. The corresponding PA rate coefficient for the given density $n_c^K \sim 5 \times 10^{10}\,\mathrm{atoms/cm^3}$ of ^{40}K atoms can be estimated to $\beta_{\mathrm{PA}} = (2.2 \pm 1.1) \times 10^{-12}\,\mathrm{cm^3/s}$.

In the second procedure, the molecule formation rate is derived from the time evolution of the atom number in the ^{6}Li-MOT after the (resonant) PA beam is switched on. The rate of change of the atom number for small times then yields the molecule formation rate. Figure 4.16 (a) depicts the atom number in the ^{6}Li-MOT as a function of the exposure time of the PA light resonant with the transition to the $\Omega = 1^{\mathrm{up}}, v = 2, J = 1$ level. The graph was recorded by measuring the minimum value of the atom number in the ^{6}Li-MOT obtained when scanning the frequency of the PA laser across the molecular resonance with different scanning speeds. The investigated molecular resonance has a width of $\sim 300\,\mathrm{MHz}$ (FWHM) (see Fig. 4.11 (b)). We thus define the ratio of that width and the scanning speed as the PA time t_{PA}. The slope of the graph at $t_{\mathrm{PA}} = 0$ yields the molecule formation rate $\sim 5 \times 10^7$ molecules/s and a PA rate coefficient of $\beta_{\mathrm{PA}} = (3.1 \pm 1.5) \times 10^{-12}\,\mathrm{cm^3/s}$, which are in accordance with those obtained in the previously described procedure. Furthermore, the graph demonstrates the importance of slowly scanning the PA laser for the optimization of the PA-induced

4.2. Experimental results

losses.

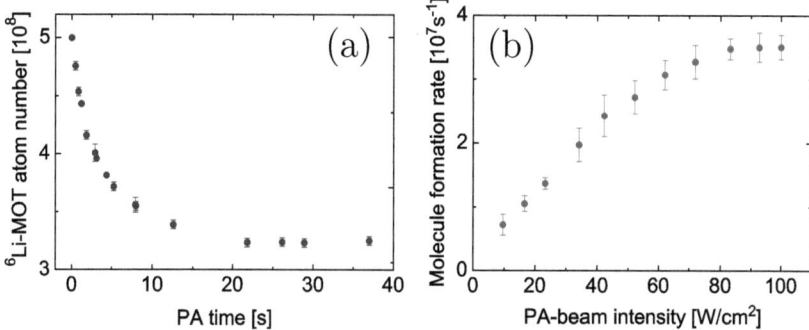

Figure 4.16: (a) Atom number in the ^{6}Li-MOT as a function of the exposure time of the resonant PA light for the resonance $\Omega = 1^{\mathrm{up}}, v = 2, J = 1$ at a PA detuning of $-14.4\,\mathrm{GHz}$ with a PA beam intensity of $100\,\mathrm{W/cm^2}$. (b) ^{6}Li^{40}K* molecule formation rate as a function of the PA beam intensity for the same molecular transition.

The obtained PA rate coefficient is larger by about a factor of two than the one found in the experiment with RbCs* [194], showing that PA rates for LiK* are much more favorable than previously expected [198], confirming the trend discussed in Ref. [199]. Using Eq. (51) of Ref. [229] we compute a coefficient of $4 \times 10^{-13}\,\mathrm{cm^3/s}$ for the transition between a continuum level of the $X^1\Sigma^+$ state at $1.2\,\mathrm{mK}$ and the singlet component of a computed 1^{up} level with $-21\,\mathrm{GHz}$ detuning. Assuming that the corresponding transition from the $a^3\Sigma^+$ state toward the triplet component of this 1^{up} level has a comparable strength, and that the $X^1\Sigma^+$ and $a^3\Sigma^+$ states are statistically populated in the initial collision, we estimate the total PA rate coefficient to $1.6 \times 10^{-12}\,\mathrm{cm^3/s}$, in agreement with our measured value.

We now compare the measured molecule formation rate for ^{6}Li^{40}K* with the one obtained simultaneously for ^{40}K$_2^*$. The highest observed trap loss in the K fluorescence due to the formation of ^{40}K$_2^*$ molecules is $\sim 12\%$ (see Fig. 4.11 (a)). Assuming a constant atom density in the ^{40}K-MOT, the rate equation for the atom number yields a ^{40}K$_2^*$ molecule formation rate of $\beta_{\mathrm{PA}} n_c^{\mathrm{K}} N_{\mathrm{K}}^{\mathrm{PA}}/2 \sim 5.3 \times 10^7\,\mathrm{s^{-1}}$, where the division by 2 accounts for the fact that two atoms are lost from the trap when a molecule is formed. The ^{40}K$_2^*$ molecule formation rate is thus comparable with the one obtained for ^{6}Li^{40}K*. It has to be noted, however, that a direct comparison of these values is difficult. First, the overlap between the PA laser and the two MOTs is not the same, as both MOTs have a different size. Second, both MOTs might have a different behavior with respect to recapture of atoms leading to a different detection efficiency for both molecules. Due to the much larger radii of the created ^{40}K$_2^*$ molecules compared to that of the ^{6}Li^{40}K* molecules, dissociation of the ^{40}K$_2^*$ molecules to free atoms is more likely. As those have a chance to be recaptured the number of actually created K$_2^*$ molecules might be underestimated. Third, the assumption about the constant atom density in the MOTs leads to different errors in the calculation of the molecule formation rate.

Figure 4.16 (b) depicts the ^{6}Li^{40}K* molecule formation rate as a function of the PA laser intensity. The graph was recorded by measuring—for different PA beam intensities—the maximum relative ^{6}Li atom loss when scanning the frequency of the PA laser across the molecular resonance. The scanning speed (of 7 MHz/s) was chosen such that the time scale of the change of the molecule formation rate is small compared to that of all loss rates appearing in Eq. (4.22) such that the molecule formation rate for each intensity could be derived from the measured ^{6}Li loss using Eq. (4.22). The curve shows the onset of a saturation. We attribute this saturation to the saturation of the molecular transition. As has been argued in Ref. [193], internal MOT processes such as optical pumping, thermalization or diffusion can be excluded as the origin of the observed behavior, since those happen on a much smaller time scale than the losses due to PA. Also, we did not observe significant broadening of the observed resonance with increasing PA laser intensity. We intend to analyze Fig. 4.16 (a) in more detail in the future with the approach described in Ref. [230] in order to deduce a saturation intensity for the investigated molecular transition. The observed onset of a saturation at reasonable laser intensities already promises efficient coherent multi-photon population transfers to the molecular rovibrational ground state.

In order to estimate the capacity of multi-photon population transfers to deeply bound levels of the electronic ground-state potential $X^1\Sigma^+$, we calculated the probabilities for decay of the created excited molecules into these states. Figure 4.17 shows the relative populations of these states after decay from two computed high-lying vibrational levels of the 1^{up} state ($v = 1$ and $v = 3$, v being counted from dissociation) calculated in the coupled state approach. The populations are given by the squared overlap integral between the radial wave functions of the computed levels with the wave functions of the $X^1\Sigma^+$ vibrational levels. The computed levels of the 1^{up} state have binding energies of -3 GHz and -81 GHz and differ from the measured ones, but their decay pattern is representative for the ones expected in the range of experimental detunings. Figure 4.17 shows that the decay rates are largest for the the most weakly bound ground-state level, but that they are still significant for the decay into deeply bound levels such as the $X^1\Sigma^+(v' = 3)$ level (v' counted from the potential bottom), which could thus be reached quite efficiently during a coherent multi-photon population transfer. The local maximum of the relative population at $v' = 3$ results from a

4.3. Conclusion

spatially coincident inner vibrational turning point of the ground and excited levels. Taking our observed PA rate of $\sim 3.5 \times 10^7\,\text{s}^{-1}$ for the $\Omega = 1^{\text{up}}, v = 2, J = 1$ level at $-14.4\,\text{GHz}$ detuning, we predict that $X^1\Sigma^+$ state molecules will be formed in the vibrational level $v' = 3$, bound by $\sim -5500\,\text{cm}^{-1}$, at a rate of $\sim 5 \times 10^4\,\text{s}^{-1}$.

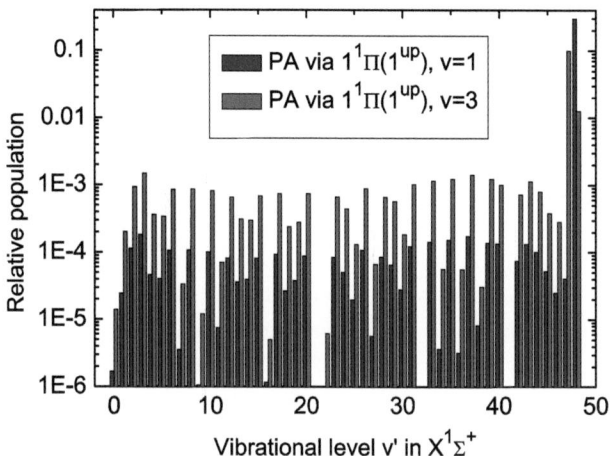

Figure 4.17: Relative populations of the different vibrational levels v' (v' being counted from the potential bottom) of the electronic ground-state potential $X^1\Sigma^+$ of LiK after decay from two computed high-lying vibrational levels of the $1^1\Pi(1^{\text{up}})$ state ($v = 1$ and $v = 3$, v being counted from dissociation). The populations are given by the squared overlap integral between the radial wave functions of the computed levels with the wave functions of the $X^1\Sigma^+$ vibrational levels.

4.3 Conclusion

In conclusion, we have demonstrated, for the first time, heteronuclear photoassociation of LiK*. We have recorded and assigned photoassociation spectra below the two lowest excited atomic asymptotes. We have derived the long-range dispersion coefficients and found a very good agreement with the theoretical predictions of Bussery et al. [166]. The measured rotational constants also agree well with our theoretical calculations based on our calculated potential curves. In particular, we have observed large formation rates for the heteronuclear $^6\text{Li}^{40}\text{K}^*$ molecules which are larger than those reported for other heteronuclear molecules in the literature [195, 196, 84, 197] except RbCs* [194] and which are comparable to those found for homonuclear ^{40}K$_2^*$. The large formation rates demonstrate that the photoassociation probability for LiK* is much more favorable than previously expected [198]. Using a theoretical model we have inferred decay rates to the deeply bound electronic ground-state vibrational level $X^1\Sigma^+(v' = 3)$ of $\sim 5 \times 10^4 \text{s}^{-1}$. These large rates promise efficient creation of rovibrational ground-state molecules via multi-photon population transfers and show that photoassociation is an attractive alternative to Feshbach resonances, since those have a very small width for ^{6}Li^{40}K and are thus difficult to control [90].

Our spectroscopic measurements represent the first measurements of binding energies of weakly bound LiK* molecules. Combining our data with data obtained in conventional molecular spectroscopy experiments for deeply bound molecules will allow the derivation of more precise potential energy curves. In particular for the excited molecular potential 1(B)$^1\Pi$, which turns into the $\Omega = 1^{\text{up}}$ potential at long range, we believe that we have recorded all of the previously undetermined vibrational level positions, such that a complete set of vibrational level positions is now available for this potential. Previously, the dissociation energy of the potential could be determined only with an uncertainty of $\pm h \times 150$ GHz [215]. We believe that the inclusion of our data will lead to an improvement of this precision of about three orders of magnitude. The gained information about the potential will allow one to find ways to efficiently transfer atoms to stable deeply bound ground-state molecules.

Many optical transfer schemes for the production of deeply bound ground-state molecules have been proposed and realized in the past years for both homo- and heteronuclear molecules [202, 98, 183]. Those can be divided into two classes, that is, single-color and multi-color transfer schemes. In single-color transfer schemes, the atoms are photoassociated into an excited molecular level which has a large probability to spontaneously decay into a deeply bound vibrational level of the electronic ground state. Sometimes, specific interaction properties lead to favorable molecular potential curves which yield high Franck-Condon probabilities for both the excitation and decay transitions, such that deeply bound ground-state molecules can be formed with reasonable efficiency. An impressive demonstration has been given for LiCs [84]: heteronuclear atom pairs could be transferred to the absolute rovibrational ground state via a single photoassociation step with an efficiency of $\sim 10^{-3}$. The disadvantage of single-color transfer schemes is that the involved spontaneous decay is an incoherent process which leads to a wide distribution of vibrational levels in the ground state and which furthermore heats the gas due to the recoil energy of the emitted photons. Multi-color transfer schemes allow the coherent population of one single level of the ground-state potential without heating of the gas. One example is the stimulated Ra-

4.3. Conclusion

man adiabatic passage (STIRAP) scheme [116, 117] mentioned in the introduction of this book (see chapter 1). It implicates two laser frequencies and requires three bound levels. An appropriate time sequence for the two laser fields allows the transfer of the populations of one molecular ground state to the other by implication of an excited molecular state, which is never populated. The STIRAP scheme cannot be applied when the continuum is the starting level [231], such that atoms have to be bound to molecules before. It has been demonstrated [118, 85, 119] that the use of Feshbach resonances for this step is very convenient, although not indispensable.

The STIRAP scheme is very promising for the achievement of a quantum degenerate gas of polar ground-state molecules. The creation of ^{6}Li^{40}K molecules by photoassociation demonstrated in this book as well as the recent realization of weakly bound ground-state ^{6}Li^{40}K molecules using Feshbach resonances provide a starting point for the application of the scheme. Our spectroscopic results will help to find pathways for an efficient STIRAP-based coherent transfer of these molecules into their rovibrational ground state. The large intrinsic electric dipole moment of deeply bound ^{6}Li^{40}K ground-state molecules would provide access to qualitatively new quantum regimes inaccessible for neutral atoms [96, 97, 98].

Chapter 5

Particle motion in rapidly oscillating potentials

5.1 Introduction

In the previous chapter we have reported on the creation of ultracold heteronuclear ^{6}Li^{40}K* molecules by photoassociation. Our results represent the first step towards the formation of polar ultracold ground-state ^{6}Li^{40}K molecules which offer a large variety of applications. In order to implement these applications, it will be necessary to have different experimental tools available which allow the control and manipulation of the molecules. The common laser-based manipulation methods, which work well for atoms, are, however, less applicable to molecules due to their complicated internal level structure. The development of manipulation methods that do not rely on the internal properties of the investigated particles is thus of great interest. In this chapter we present and study a novel manipulation method which is applicable to all kinds of particles, such as, e.g., polar ^{6}Li^{40}K molecules. This method consists in trapping the particles in a rapidly oscillating potential (ROP) and to induce an instantaneous change of phase of the trapping potential (a phase hop). We show that the particle's motion can thus be significantly manipulated in a controlled fashion. In the following we present a general theoretical treatment of particle motion in rapidly oscillating potentials under the influence of phase hops. We here mainly focus on the quantum regime, since we are interested in describing ultracold particles.

Potentials which oscillate rapidly relative to the motion of particles inside them are widely used to trap charged and neutral particles for various applications. The reasons for this are manifold. Most notably, rapidly oscillating potentials can allow trapping in cases where static potentials cannot. This is due to a property of ROPs which is termed *dynamic stabilization*, which was first demonstrated in 1951 by Kapitza [232], utilizing an inverted pendulum whose suspension point was forced to oscillate vertically. He showed that for high enough frequencies and large enough amplitudes of this oscillation the upwards position of the pendulum becomes stable. Dynamic stabilization is the basic principle of the so-called dynamic traps. Well-known paradigms are the Paul trap for charged particles [233, 234, 235] and the electro- and magnetodynamic traps for high-field seeking polar molecules [236, 237] and neutral atoms [238, 239, 240, 241].

Another reason for the wide use of ROPs is that they allow the realization of complicated trap geometries. Prime examples are the TOP-trap [152, 242, 243, 244, 245, 246, 247], the optical billiard traps [248, 249] and rapidly scanning optical tweezers [250, 251] for ultracold neutral atoms. The latter are also used in Biophysics to trap and investigate dielectric particles such as polymers and living cells [252, 253].

Finally, ROPs are attractive for use, because the description of the motion of particles inside them—as compared to other time-varying potentials—is very simple. This is because the particles' *mean*-motion (averaged over the ROP's fast oscillations) is in both the classical and the quantum regime to a good approximation determined by a static *effective* potential [254, 255, 256, 257].

Our preliminary calculations for the classical regime [124] show that in ROPs with a vanishing time-average—such as *e.g.*, the Paul trap—the mean motion of trapped particles is strongly coupled to the phase of the ROP. Consequently, the particles' mean motion can be appreciably manipulated by changing the phase of the ROP. For the Paul trap, a phase hop can change the mean-energy of a trapped classical particle (that is not constantly at rest) by a factor which can take any value between 0.1 and 9.9, independent of the particle's mean-energy [124], thus offering a powerful tool for particle manipulation.

Here we are mainly interested in *quantum* particles trapped in ROPs, since those offer many applications [234, 235, 152, 245, 246, 248, 249, 250, 251]. It is not clear how a phase hop affects quantum particles: in the Paul trap, a classical particle which does not move is not affected by a phase hop [124]. Thus, a quantum particle in, *e.g.*, the ground state of the effective trapping potential might also not or only weakly be affected by a phase hop. Furthermore, the ability to dissipate up to 90% of a particle's mean-energy in a Paul trap by a single phase hop, which is given in the classical regime independent of the particle's energy, can not generally hold in the quantum regime.

We have derived an independent quantum mechanical treatment of the effect of phase hops on a particle trapped by a ROP of arbitrary shape. It is shown here by both analytical and numerical calculations that a phase hop can strongly influence the particle's mean motion, even if it is in the ground state of the effective trapping potential. The results of this work have lead to a publication [125] (see Appendix D)

5.1. Introduction

and are presented in this chapter.

Recently, the proposed method has been implemented experimentally with an ensemble of quantum particles which form a Bose-Einstein condensate (BEC) trapped in a time-averaged orbiting potential (TOP) [126]. In this experiment, the BEC was prepared in a static Ioffe-Pritchard (IP) magnetic trap and subsequently transferred to the TOP to allow the realization of complicated trap geometries (as *e.g.* a double-well potential). The sudden replacement of the IP trap by the TOP induces an undesired oscillating mean motion of the BEC. It was shown that these oscillations could be nearly entirely quenched by applying a single phase hop to the TOP potential, supporting the statements which are presented in this chapter. Furthermore, the experiment demonstrated that despite the imperfections given in an experiment, phase hops can be induced with a high precision.

Throughout this chapter we use one-dimensional formalism, but the results presented here can be directly applied to two and three dimensions in cases where the motion is separable [254]. For high-dimensional oscillating potentials where the particle's motion in not separable and in particular chaotic, new phenomena may occur that are not discussed here [249]. In Ref. [126], the effect of phase hops in a TOP trap was generalized to three dimensions.

The remainder of this chapter is organized as follows. In Sec. 5.2 we give an overview of the classical treatment of the motion of a particle in a ROP and the effect of a phase hop. We show that in the limit of large driving frequencies of the ROP the time-dependent equation of motion can be replaced by a time-independent one with a static effective potential. We show that inducing a phase hop does not change this time-independent equation of motion. Analytical calculations show that a phase hop can significantly influence the particle's mean motion. We demonstrate that the effect of a phase hop can be visualized as being the result of a collision with an imaginary particle. In Sec. 5.3 we address the corresponding quantum mechanical problem. Also here, the equation of motion of the quantum particle can be replaced by a time-independent one. The particle's mean motion is found to possess stationary mean-motion states. We show that a phase hop affects these states in a significantly different way than it affects classical states: whereas a phase hop can be always induced such that the energy of a classical particle can be decreased, its effect on a quantum particle in a stationary mean-motion state always consists in increasing the particle's mean energy. Further we calculate the transition probabilities between stationary mean-motion states which are induced by a phase hop. Finally the consistency of the quantum mechanical results with the corresponding classical results is verified using coherent states. For the model potential of rapidly scanning optical tweezers, we find that a phase hop does not destroy the coherence of a coherent state, showing that phase hops could be used, in principal, to decelerate a classical particle until it occupies the mean-motion ground state. For the model potential of the Paul trap, however, a phase hop is found to destroy the coherence of coherent states.

5.2 Classical motion in a rapidly oscillating potential

5.2.1 Time-independent description

The time-independent description of the motion of a particle in a rapidly oscillating potential has been studied several times in the past [254, 255, 256, 257]. It has been shown that the mean motion of a classical particle which is moving in a ROP is approximately governed by an equation of motion with a time-independent effective potential. The derivation is given in the following. The motion of a particle in a rapidly oscillating one-dimensional potential $V(x,t)$ is described by Newton's equation

$$m\ddot{x} = -V_0'(x) - V_1'(x,\omega t), \qquad (5.1)$$

(primes denote derivatives with respect to x) with ω denoting the driving frequency of the ROP. Here, the last two terms represent a separation of V into a time-averaged part V_0 and an oscillating part V_1 with a vanishing period average. Two experimentally relevant examples for the considered type of potentials are

$$V^{\mathrm{PT}}(x,\omega t) = \frac{1}{2}m\omega_{\mathrm{osc}}^2 x^2 \cos\omega t, \qquad (5.2)$$

$$V^{\mathrm{OT}}(x,\omega t) = \frac{1}{2}m\omega_{\mathrm{osc}}^2 (x - x_0 \cos\omega t)^2, \qquad (5.3)$$

(shown in Fig. 5.1) for which $V_0^{\mathrm{PT}}(x) \equiv \overline{V^{\mathrm{PT}}(x,\omega t)} = 0$ (the overbar denotes the time-average over one period) and $V_0^{\mathrm{OT}}(x) \equiv \overline{V^{\mathrm{OT}}(x,\omega t)} = m\omega_{\mathrm{osc}}^2(x^2 - x_0^2/2)/2$. These potentials are model-potentials for the Paul-trap (5.2) and for rapidly scanning optical tweezers (5.3), respectively.

The potential V is assumed to oscillate rapidly, that is, its period is inferior to the characteristic time of the system's dynamics. One observes that at these large driving frequencies, the particle's motion separates into two parts that evolve on the different time scales t and $\tau \equiv \omega t$ [255, 256, 257]. The particle's motion can thus be written as

$$x(t) = X(t) + \xi(X,\tau), \qquad (5.4)$$

5.2. Classical motion in a rapidly oscillating potential

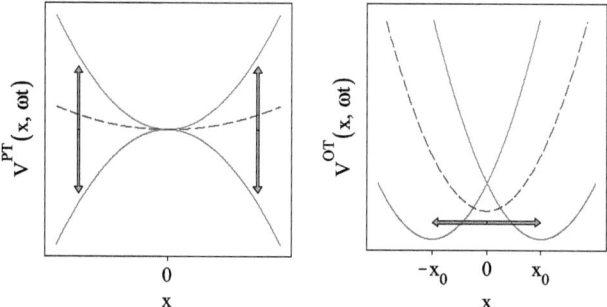

Figure 5.1: Model-potentials for the Paul-trap (5.2) (left) and for spatially oscillating optical tweezers (5.3) (right). Solid (red) curves: time-dependent potential, dashed (blue) curves: time-independent effective potential.

where $\xi(X, \tau)$ is a time-periodic function with the same period as V, which has a small amplitude and a vanishing period average. Accordingly, it is termed *micromotion*. $X(t)$ is a slowly varying function of time, which is the solution of a Newton equation with a time-independent potential and which describes the particle's mean motion. The knowledge of this Newton equation allows one to obtain an intuition for the particle's mean motion as it does not depend on time. An approximation for this equation can be obtained in the following two steps [232, 254, 258]. First, Eq. (5.4) is substituted into Eq. (5.1) and the potential terms are expanded in powers of ξ as far as the first order terms. Second, ξ is chosen as

$$\xi(X,\tau) = -\frac{1}{m\omega^2} \int^{(2)\tau} [V_1'(X,\tau)], \tag{5.5}$$

(which implies $\overline{\xi(X,\tau)} = 0$) and the resulting equation is averaged in time over one period of V. The integral $\int^\tau [f(\tau)]$ of a periodic function $f(\tau)$ is defined as the antiderivative of $f(\tau)$ which has a vanishing period average [259]: $\int^\tau [f(\tau)] \equiv \int_0^\tau f(\tau')d\tau' - \overline{\int_0^\tau f(\tau')d\tau'}$ [1]. It can be applied repeatedly. Its multiple application (j times) is denoted by $\int^{(j)\tau}[f(\tau)]$ and its evaluation at the point τ_0 by $[\int^\tau[\cdots]]_{\tau_0}$. The choice (5.5) of ξ implies that the time-dependent potential term $-V_1'(X,\tau)$ in Eq. (5.1) is canceled by the term $m\ddot{\xi}$ such that only time-dependent terms with a small amplitude remain. The averaging finally results in

$$m\ddot{X} = -V_{\text{eff}}'(X), \tag{5.6}$$

---

[1] For example, it is $\int^\tau[\cos(\tau)] = \sin(\tau)$, $\int^{(2)\tau}[\cos(\tau)] = -\cos(\tau)$.

with the effective potential

$$V_{\text{eff}}(X) = V_0(X) + \frac{1}{2m\omega^2}\overline{\left(\int^\tau [V_1'(X,\tau)]\right)^2} = V_0(X) + \frac{1}{2}m\overline{\dot{\xi}(X,\tau)^2} \qquad (5.7)$$

(dots denote derivatives with respect to t). The second term on the right hand side (rhs) of Eq. (5.7) can be interpreted as the mean kinetic energy stored in the micromotion. For the above examples one has $V_{\text{eff}}^{\text{PT}}(x) = m\Omega^2 x^2/2$, with $\Omega \equiv \omega_{\text{osc}}^2/(\omega\sqrt{2})$, and $V_{\text{eff}}^{\text{OT}}(x) = m\omega_{\text{osc}}^2 x^2/2 + c$, with an x-independent constant $c = m\omega_{\text{osc}}^2 x_0^2 \left(1 + \omega_{\text{osc}}^2/\omega^2\right)/4$ (see Fig. 5.1). In both model potentials the effective potential equals a harmonic oscillator potential, which is capable to trap the particle. For the Paul trap potential (5.2) this is surprising as it has a vanishing time average [2].

We have seen that the particle's mean motion inside a ROP is governed by a Newton equation with a time-independent effective potential. As a result, the mean energy of the particle in a ROP is conserved. In particular we have seen, that the effective potential can have a local minimum even in cases where the time-average of the ROP vanishes.

5.2.2 Coupling between the mean motion and the potential's phase

In the following we investigate how the particle's mean motion is related to the phase φ of the ROP $V(x,\omega t+\varphi)$. As the effective potential (5.7) consists only of period-averaged terms, it is independent of the phase of the ROP and thus also the equation of motion (5.6) for the particle's mean motion. In the first instance one might thus naively expect that the mean motion does not depend on the potential's phase. However, the motion of a particle not only depends on its equation of motion but also on its initial conditions. According to Eq. (5.4) the initial conditions $X(0)$ and $\dot{X}(0)$ for the particle's mean motion are related to those of the real motion $x(0)$ and $\dot{x}(0)$ via

$$X(0) = x(0) - \xi(0) = x(0) + \frac{1}{m\omega^2}\left[\int^{(2)\tau} [V_1'(X,\tau+\varphi)]\right]_{\tau=0}, \qquad (5.8)$$

$$\dot{X}(0) = \dot{x}(0) - \dot{\xi}(0) = \dot{x}(0) + \frac{1}{m\omega}\left[\int^\tau [V_1'(X,\tau+\varphi)]\right]_{\tau=0}. \qquad (5.9)$$

---

[2] Note that for the Paul trap potential the oscillation frequency Ω of the effective potential is much smaller than the maximum instantaneous oscillation frequency ω_{osc} of the oscillating potential. For the optical tweezers potential (5.3) both these oscillation frequencies are the same.

5.2. Classical motion in a rapidly oscillating potential

Since the micromotion ξ depends on the phase φ, also $X(0)$ and $\dot{X}(0)$ do and thus the resulting mean motion of the particle. In the considered limit of very large driving frequencies both, $\xi(0)$ and $\dot{\xi}(0)$, become small as they scale with ω^{-2} and ω^{-1}, which might lead one to think, however, that the initial condition transformations (5.8) and (5.9), which lead to the phase dependence of the particle's mean motion, are insignificant. But for certain cases, the transformations can not be neglected [124]. This can be seen from the following argument. Consider a particle that is trapped inside a ROP with a vanishing average. According to Eq. (5.7), the potential energy of the particle's mean motion is then proportional to the mean kinetic energy stored in its micromotion. For a trapped particle the potential energy of the particle's mean motion can be larger than the kinetic energy of the mean motion and thus, the kinetic energy stored in the micromotion is not negligible with respect to the kinetic energy of the particle's mean motion. Consequently, $\dot{\xi}(0)$ cannot be neglected compared to $\dot{X}(0)$ in Eq. (5.9) even for arbitrarily large driving frequencies ω [3]. For particles trapped in ROPs with a vanishing average, the mean motion is thus strongly coupled to the phase of the potential. This strong coupling can also exist for ROPs with a non-vanishing average.

A numerical demonstration of the significant dependence of the particle's mean motion on the potential's phase is depicted in Fig. 5.2 for the two model potentials $V^{\text{PT}}(x,\omega t+\varphi)$ (5.2) and $V^{\text{OT}}(x,\omega t+\varphi)$ (5.3). The trajectories in the figure are obtained by numerically solving the time-dependent Newton equation (5.1) for given initial conditions and very large driving frequencies ω. The trajectories appreciably differ from each other in phase and amplitude as expected from the derived theory. Figure 5.2 shows as well, that all trajectories appear to be generated from the same time-independent effective potential, which thus does not depend on the phase φ. Since the trajectories for different φ are not the same although their initial conditions are, this verifies that the trajectories are determined in addition by a non-negligible transformation of their initial conditions which depends on φ. The shown trajectories are almost indistinguishable from their theoretical predictions calculated from equations (5.6) and (5.9), which are therefore not shown in the figure.

5.2.3 The effect of a phase hop

In the following we show, that the strong coupling between the particle's mean motion and the phase of the ROP can be utilized to significantly manipulate the particle's mean motion by modifying the phase of the potential. We focus here on the simple

---

[3]The micromotion position ξ however, can be neglected with respect to the mean position X for large ω, since this was the basic assumption for the time-independent description of the particle's mean motion.

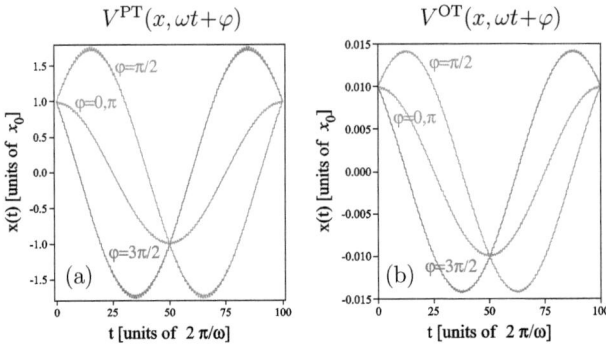

Figure 5.2: Trajectories of a particle inside (a) the Paul-trap potential $V^{\text{PT}}(x,\omega t+\varphi)$ (5.2) with initial conditions $x(0) = x_0$ and $\dot{x}(0) = 0$ and the driving frequency $\omega = 100\Omega$ and (b) the oscillating tweezers potential $V^{\text{OT}}(x,\omega t+\varphi)$ (5.3) with initial conditions $x(0) = 0.01x_0$ and $\dot{x}(0) = 0$ and the driving frequency $\omega = 100\omega_{\text{osc}}$ for different phases φ of the potential. The trajectories are obtained from a numerical solution of the time-dependent Newton equation (5.1).

case, where the phase of the potential $V(x,\tau)$ is, at a time t_{ph}, *instantaneously*[4] changed from $\tau_{\text{ph}} \equiv \omega t_{\text{ph}}$ to $\tau_{\text{ph}}+\Delta\varphi$ (a phase hop). In such a case the particle is, for $t > t_{\text{ph}}$, moving in the ROP $V_{\text{new}}(x,\tau) \equiv V(x,\tau+\Delta\varphi)$ and its mean motion is governed by Eq. (5.6) with an effective potential $V_{\text{eff}}^{\text{new}}(X)$. As the effective potential of a ROP consists only of period-averaged terms (Eq. (5.7)), it is independent of the phase of the ROP, implying $V_{\text{eff}}^{\text{new}}(X) = V_{\text{eff}}(X)$. Thus the equation of motion for the particle's mean motion remains unchanged. However, the mean motion itself changes due to the continuity of the particle's real motion $x(t)$ at $t=t_{\text{ph}}$: for $t \neq t_{\text{ph}}$ the latter is the sum of the micromotion and the mean motion (Eq. (5.4)). As the micromotion depends on the phase of V (Eq. (5.5)), it changes instantaneously and thus involves a corresponding change of the mean motion. As we have seen in Sec. 5.2.2, the micromotion velocity and thus its change, which can be induced by a phase hop, is non-negligible. As a result, the mean motion of a particle in a ROP can be significantly manipulated by inducing a phase hop.

We now calculate the instantaneous changes of the particle's mean-motion variables, which are induced by a phase hop. The continuity condition for the particle's real

---

[4] on a time scale smaller than the period of V

5.2. Classical motion in a rapidly oscillating potential

motion $x(t)$ implies

$$\Delta X(t_{\rm ph}) = X^{\rm new}(t_{\rm ph}) - X(t_{\rm ph}) = x^{\rm new}(t_{\rm ph}) - x(t_{\rm ph}) = 0, \qquad (5.10)$$

$$\Delta \dot{X}(t_{\rm ph}) = \dot{X}^{\rm new}(t_{\rm ph}) - \dot{X}(t_{\rm ph}) = \dot{\xi}(t_{\rm ph}) - \dot{\xi}^{\rm new}(t_{\rm ph})$$

$$= \frac{1}{m\omega} \int_0^{\omega t_{\rm ph}} [V_1'(X(t_{\rm ph}), \tau + \Delta\varphi) - V_1'(X(t_{\rm ph}), \tau)]\, d\tau \qquad (5.11)$$

(the notation $\lim_{t \to t_{\rm ph}, t < t_{\rm ph}} X(x,t) \equiv X(x, t_{\rm ph})$ and $\lim_{t \to t_{\rm ph}, t > t_{\rm ph}} X^{\rm new}(x,t) \equiv X^{\rm new}(x, t_{\rm ph})$ is used). Equation. (5.11) shows that $\Delta \dot{X}(t_{\rm ph})$ equals the change of velocity of the particle's micromotion, which is taking place at the same time, demonstrating that the phase hop causes a momentum transfer between the micromotion and the mean motion. The fact that the phase hop can change the particle's mean momentum instantaneously, but not its mean position, shows that its effect on the particle's mean motion can be visualized as being the result of a collision with an imaginary particle[5]. The particle's mean energy, which is conserved before and after the phase hop, changes by

$$\Delta E(t_{\rm ph}) = m\dot{X}(t_{\rm ph})\Delta \dot{X}(t_{\rm ph}) + \frac{1}{2} m [\Delta \dot{X}(t_{\rm ph})]^2. \qquad (5.12)$$

This change can be negative as well as positive.

For the above model potentials the changes of the mean-motion variables take simple forms. For a particle with (non-zero) total mean energy E and mean kinetic energy αE (with $0 \leq \alpha \leq 1$) one obtains for the Paul trap (5.2): $\Delta \dot{X} = X \delta \omega_{\rm osc}^2 / \omega$ with $\delta \equiv \sin(\tau_{\rm ph} + \Delta\varphi) - \sin(\tau_{\rm ph})$, and $\Delta E/E = \pm \sqrt{8\alpha(1-\alpha)}\delta + 2(1-\alpha)\delta^2$, whereas the $\pm$-sign accounts for the possible signs of the product $\dot{X}(t_{\rm ph}) X(t_{\rm ph})$. Thus, the relative change of the particle's mean energy is independent of its mean energy itself and of the driving frequency ω. It can take any values between $-0.9 \lesssim \Delta E/E \lesssim 9$, whereas the extreme values are taken if $\alpha \approx 0.91$, $\dot{X}(t_{\rm ph}), X(t_{\rm ph}) > 0$. It then is $\Delta E/E \approx -0.9$ for $\delta = -2$ and $\Delta E/E \approx 9$ for $\delta = 2$. For the special case $E = 0$, it is $X(t) = 0$ and thus $\xi(t) = 0$, such that in this case the particle is not affected by the phase hop.

For the oscillating tweezers potential (5.3) one has $\Delta \dot{X} = x_0 \delta \omega_{\rm osc}^2 / \omega$. Note, that $\Delta \dot{X}$ is independent of the particle's position. This means, that if an ensemble of non-interacting particles moved inside the oscillating tweezers potential, all particles would, in case of a phase hop, change their velocity by the same amount. A phase hop thus simply induces a Galilean transformation in this system. The absolute change in the particle's mean energy is given by $\Delta E = \pm \sqrt{2m\alpha E} x_0 \delta \omega_{\rm osc}^2 / \omega + x_0^2 \delta^2 \omega_{\rm osc}^4 / (2\omega^2)$. The

---

[5] This analogy only approximately holds for the particle's real motion, since the particle's micromotion is affected differently by a phase hop than by a collision: a phase hop induces an instantaneous change of the phase of the micromotion, whereas a collision would not [260].

resulting relative change of the mean energy depends on the mean energy itself and on the driving frequency ω. It is bigger for smaller mean energies and smaller driving frequencies. Nonetheless, it can still be very large even for the considered limit of very large driving frequencies: for a particle with mean energy $E = 0.01 m\omega_{\mathrm{osc}}^2 x_0^2$ and for the driving frequency $\omega = 100\omega_{\mathrm{osc}}$ one obtains $-0.26 \lesssim \Delta E/E \lesssim 0.3$. Here the extreme values are taken if $\alpha = 1$ (i.e. $X(t_{\mathrm{ph}}) = 0$) and $\dot{X}(t_{\mathrm{ph}}) > 0$. It then is $\Delta E/E \approx -0.26$ for $\delta = -2$ and $\Delta E/E \approx 0.3$ for $\delta = 2$.

A numerical demonstration of the significant impact that a phase hop can have on the particle's mean motion is depicted in Fig. 5.3 for the Paul trap potential $V^{\mathrm{PT}}(x,\tau)$ (5.2). The trajectories in the figure are obtained from the time-dependent Newton equation (5.1) for the large driving frequency $\omega = 100\Omega$. Figure 5.3 (a) demonstrates that the particle's mean energy can be significantly increased by a phase hop. Figure 5.3 (b) demonstrates that the particle's mean energy can also be significantly decreased by a phase hop. In particular it demonstrates, that several phase hops can have accumulating effects on the particle's mean motion, even though the time-dependent Newton equation (5.1) is transformed into its initial one. Since the change of the particle's mean energy in the Paul-trap potential does not depend on the particle's mean-energy itself, phase hops can be used to decrease the particle's mean energy exponentially fast (and within one period of the particle's mean motion, see Fig. 5.3 (b)). Figure 5.3 (c) demonstrates that the effect of a phase hop can be canceled by a second one. This is in general always possible in any ROP which has a harmonic effective potential. Finally, Fig. 5.3 (d) demonstrates that a series of two phase hops can also be used in order to change the phase of the particle's mean motion without changing its energy.

We have seen, that the strong coupling between the mean motion of a particle inside a ROP and the phase of the ROP can be used to efficiently manipulate the particle's mean motion by instantaneously changing the phase of the ROP (a phase hop). We found that a phase hop does not change the effective potential for the particle's mean motion and that it can be visualized as being the result of a collision with an imaginary particle, which can be controlled. For ROPs with a vanishing time-average, the change of the particle's mean energy due to a phase hop was found to be independent of the particle's mean-energy itself and of the potential's driving frequency. We now turn to the question how a phase hop affects a quantum particle. Obviously, the mean energy of a particle cannot be infinitely decreased, since a trapped quantum particle has a finite mean energy due to Heisenberg's uncertainty relation. In order to answer this question, we give, in the subsequent section, an independent treatment of the effect of phase hops on quantum particles.

5.2. Classical motion in a rapidly oscillating potential

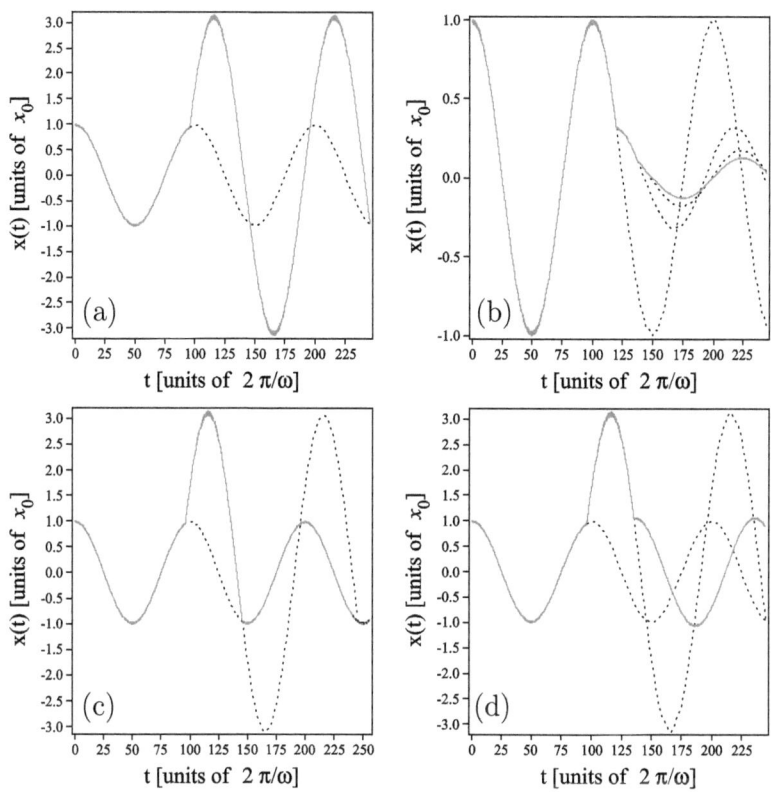

Figure 5.3: Examples of trajectories of a particle inside the Paul-trap potential $V^{\mathrm{PT}}(x,\omega t)$ (5.2) when one or more phase hops of the size $\Delta\varphi = \pi$ are induced at different times. Solid (turquoise) lines: real trajectory, dotted (black) lines: imaginary continuations of trajectory when no phase hops were induced. (a) One phase hop induced at $t_{\mathrm{ph}} = 95.75(2\pi/\omega)$, leading to an increase of the particle's mean energy of a factor of approximately 9. (b) Three phase hops induced at $t_{\mathrm{ph}}^{(1)} = 119.75(2\pi/\omega)$, $t_{\mathrm{ph}}^{(2)} = 137.5(2\pi/\omega)$, $t_{\mathrm{ph}}^{(3)} = 145.25(2\pi/\omega)$, leading to a decrease of the particle's mean energy of a factor of approximately 150. (c) Two phase hops induced at $t_{\mathrm{ph}}^{(1)} = 95.75(2\pi/\omega)$, $t_{\mathrm{ph}}^{(2)} = 145.75(2\pi/\omega)$, where the effect of the first phase hop is canceled by the second. (d) Two phase hops induced at $t_{\mathrm{ph}}^{(1)} = 95.75(2\pi/\omega)$, $t_{\mathrm{ph}}^{(2)} = 135.75(2\pi/\omega)$, where the net effect of the two phase hops consists in changing the phase of the particle's mean motion without changing its energy. The trajectories are obtained from a numerical solution of the time-dependent Newton equation (5.1) with initial conditions $x(0)=x_0$, $\dot{x}(0) = 0$ and with $\omega = 100\Omega$.

5.3 Quantum motion in a rapidly oscillating potential

5.3.1 Time-independent description

The motion of a quantum particle in a rapidly oscillating potential has been studied previously [255, 256, 257]. It has been shown that the mean motion of the particle is governed by a Schrödinger equation with a time-independent Hamiltonian, which is expected from the classical results presented in the previous section. In the following we derive this effective Hamiltonian. The Schrödinger equation for a quantum particle in a ROP $V(x,\omega t)$ reads

$$i\hbar\frac{\partial}{\partial t}\psi(x,t) = \hat{H}(x,\omega t)\psi(x,t), \qquad (5.13)$$

with the time-dependent and time-periodic Hamiltonian

$$\hat{H}(x,\tau) = -\frac{\hbar^2}{2m}\frac{\partial^2}{\partial x^2} + V_0(x) + V_1(x,\tau), \qquad (5.14)$$

with $V_0(x) = \overline{V(x,\tau)}$ and $\overline{V_1(x,\tau)} = 0$. If the frequency ω is very large compared to the other time scales of the system, one expects that the potential term $V_1(x,\tau)$ in the Hamiltonian only has little effect on the particle's mean motion during one period, since its period-average is zero. In order to understand its effect on the particle's wave function $\psi(x,t)$, suppose that the Hamiltonian in Eq. (5.13) consisted only of the term $V_1(x,\tau)$. Then, the solution of Eq. (5.13) would be

$$\psi(x,t) = e^{-i\int V_1(x,\omega t)\,\mathrm{d}t/\hbar}\,\psi(x,0). \qquad (5.15)$$

Thus, the dominant effect of the potential term $V_1(x,\tau)$ is to add an oscillating phase factor to the wave function $\psi(x,t)$. This suggests that for large driving frequencies the solutions of Eq. (5.13) approximately have the functional form [255, 256, 257]

$$\psi(x,t) \approx e^{-iF(x,\omega t)}\phi(x,t), \qquad (5.16)$$

where F is a time-periodic function with the same period as $V_1(x,\tau)$ and ϕ is a slowly varying function of time, which is the solution of a Schrödinger equation with a time-*independent* effective Hamiltonian and which describes the particle's mean motion. This effective Hamiltonian can be obtained as follows. Substituting Eq. (5.16) into

5.3. Quantum motion in a rapidly oscillating potential

Eq. (5.13) and multiplying both sides by $e^{iF(x,\tau)}$ yields

$$\begin{aligned}i\hbar\frac{\partial}{\partial t}\phi(x,t) &= \left(-\frac{\hbar^2}{2m}\frac{\partial^2}{\partial x^2}+V_0(x)\right)\phi(x,t)\\&+\frac{\hbar^2}{2m}\left(iF''(x,\tau)+2iF'(x,\tau)\frac{\partial}{\partial x}+F'(x,\tau)^2\right)\phi(x,t)\\&+\left(V_1(x,\tau)-\hbar\omega\left(\frac{\partial}{\partial \tau}F(x,\tau)\right)\right)\phi(x,t).\end{aligned} \quad (5.17)$$

Choosing

$$F(x,\tau) = \frac{1}{\hbar\omega}\left(\int_0^\tau V_1(x,\tau')d\tau' - \overline{\int_0^\tau V_1(x,\tau')d\tau'}\right) \equiv \frac{1}{\hbar\omega}\int^\tau [V_1(x,\tau)], \quad (5.18)$$

precisely cancels the (possibly large) oscillating potential term $V_1(x,\tau)$ in Eq. (5.17) and only yields time-dependent terms with a small amplitude (that scale with ω^{-1} or ω^{-2}). Further, Eq. (5.18) implies $\overline{F(x,\tau)} = 0$, such that, since $\phi(x,t)$ is a slowly varying function of time, averaging of Eq. (5.17) in time over one period of the potential's oscillation results in an ordinary Schrödinger equation for $\phi(x,t)$,

$$i\hbar\frac{\partial}{\partial t}\phi(x,t) = \hat{H}_{\text{eff}}(x)\phi(x,t), \quad (5.19)$$

with the time-independent *effective* Hamiltonian

$$\hat{H}_{\text{eff}}(x) = -\frac{\hbar^2}{2m}\frac{\partial^2}{\partial x^2} + V_{\text{eff}}(x), \quad (5.20)$$

which contains an effective potential given by

$$V_{\text{eff}}(x) = V_0(x) + \frac{\hbar^2}{2m}\overline{F'(x,\tau)^2} = V_0(x) + \frac{1}{2m\omega^2}\overline{\left(\int^\tau [V_1'(x,\tau)]\right)^2}. \quad (5.21)$$

This effective potential is identical to the one for the classical case (see Eq. (5.7)). Since the effective Hamiltonian $\hat{H}_{\text{eff}}$ is time-independent, the mean-motion wave function $\phi(x,t)$ is explicitly known for all times,

$$\phi(x,t) = e^{-i\hat{H}_{\text{eff}}(x)t/\hbar}e^{iF(x,0)}\psi(x,0), \quad (5.22)$$

where $\psi(x,0)$ is the particle's state at the initial time $t=0$. For the two model potentials $V^{\text{PT}}(x,\omega t)$ (5.2) and $V^{\text{OT}}(x,\omega t)$ (5.3), the function F takes the form $F^{\text{PT}}(x,\tau) = m\Omega x^2 \sin\tau/(\hbar\sqrt{2})$ and $F^{\text{OT}}(x,\tau) = m\omega_{\text{osc}}^2(4xx_0\sin\tau + x_0^2\sin\tau\cos\tau)/(4\hbar\omega)$, respectively. $\phi(x,t)$ indeed approximately describes the particle's mean motion, since it is

$$\begin{aligned}\overline{\langle x \rangle} &\equiv \overline{\langle \psi(x,t)|\hat{x}|\psi(x,t)\rangle} \approx \overline{\langle \phi(x,t)|e^{iF(x,\tau)}\hat{x}e^{-iF(x,\tau)}|\phi(x,t)\rangle} \\ &= \langle \phi(x,t)|\hat{x}|\phi(x,t)\rangle, \end{aligned} \quad (5.23)$$

$$\begin{aligned}\overline{\langle p \rangle} &\equiv \overline{\langle \psi(x,t)|\hat{p}|\psi(x,t)\rangle} \approx \overline{\langle \phi(x,t)|e^{iF(x,\tau)}\hat{p}e^{-iF(x,\tau)}|\phi(x,t)\rangle} \\ &= \overline{\langle \phi(x,t)|\hat{p} - \hbar F'(x,\tau)|\phi(x,t)\rangle} = \langle \phi(x,t)|\hat{p}|\phi(x,t)\rangle, \end{aligned} \quad (5.24)$$

$$\overline{\langle E \rangle} \equiv \overline{\langle \psi(x,t)|\hat{H}(x,\tau)|\psi(x,t)\rangle} \approx \langle \phi(x,t)|\hat{H}_{\text{eff}}(x)|\phi(x,t)\rangle, \quad (5.25)$$

where $\hat{x} = x$ and $\hat{p} = -i\hbar\,\partial/\partial x$ denote the position and momentum operator, respectively. $\overline{\langle E \rangle}$ is referred to as the particle's mean energy. The difference between ϕ and ψ is approximately given by the oscillating phase factor e^{-iF} (Eq. (5.16)), which has a small amplitude as F scales with ω^{-1} (Eq. (5.18)), and thus describes a *micro*motion [234, 235], since it is $\langle\psi|\hat{x}|\psi\rangle-\langle\phi|\hat{x}|\phi\rangle\approx 0$ and $\langle\psi|\hat{p}|\psi\rangle-\langle\phi|\hat{p}|\phi\rangle\approx -\hbar\langle\phi|F'|\phi\rangle$. In Ref. [257], a more general ansatz than the one given by Eq. (5.16) is made, where $F(x,\tau)$ is considered to be a Hermitian operator rather than simply a function, resulting in corrections to F and $\hat{H}_{\text{eff}}$, which scale with a higher power of ω^{-1} than the terms calculated here (see Eqs. (5.18) and (5.20)). In the following, ω is assumed to be sufficiently large such that these correction terms can be neglected.

Being governed by a Schrödinger equation with a time-independent Hamiltonian, a trapped particle's mean motion possesses stationary states (the *stationary mean-motion states*). For the two model potentials $V^{\text{PT}}(x,\omega t)$ (5.2) and $V^{\text{OT}}(x,\omega t)$ (5.3), the stationary mean-motion states and their energies are explicitly known, since for these potentials the effective Hamiltonian equals that of a harmonic oscillator. For the Paul trap potential, the stationary mean motion states $\phi_n^{\text{PT}}(x)$ read

$$\phi_n^{\text{PT}}(x) = \left(\frac{m\Omega}{\pi\hbar}\right)^{\frac{1}{4}} \frac{1}{\sqrt{2^n n!}} H_n\left(\sqrt{\frac{m\Omega}{\hbar}}x\right) e^{-\frac{1}{2}\frac{m\Omega}{\hbar}x^2}, \quad (5.26)$$

where H_n denotes the n-th Hermite polynomial. Their energies, the eigenvalues of $\hat{H}_{\text{eff}}^{\text{PT}}$, are given by

$$E_n^{\text{PT}} = \hbar\Omega\left(n + \frac{1}{2}\right), \quad (5.27)$$

and scale with ω^{-1}. For the oscillating tweezers potential, the stationary mean-motion states $\phi_n^{\rm OT}$ read

$$\phi_n^{\rm OT}(x) = \left(\frac{m\omega_{\rm osc}}{\pi\hbar}\right)^{\frac{1}{4}} \frac{1}{\sqrt{2^n n!}} H_n\left(\sqrt{\frac{m\omega_{\rm osc}}{\hbar}}x\right) e^{-\frac{1}{2}\frac{m\omega_{\rm osc}}{\hbar}x^2}. \tag{5.28}$$

Their energies, the eigenvalues of $\hat{H}_{\rm eff}^{\rm OT}$, are given by

$$E_n^{\rm OT} = \hbar\omega_{\rm osc}\left(n + \frac{1}{2}\right), \tag{5.29}$$

and scale with ω^0.

Connection with Floquet's theory

The motion of a quantum particle in a ROP is governed by a Schrödinger equation with a time-periodic Hamiltonian. Such systems can in general be treated using Floquet theory [261, 262], which is not restricted to large driving frequencies. The Floquet theorem states, that the propagator of any state $\psi(x, t_0)$, given at the initial time t_0, can be written in the form

$$\hat{U}(t, t_0) = \hat{P}(t, t_0)\, e^{-i(t-t_0)\hat{G}(t_0)/\hbar} \tag{5.30}$$

with the time-independent, Hermitian, so-called *quasienergy operator* $\hat{G}(t_0)$ and a unitary time-periodic operator $\hat{P}(t, t_0)$, which obeys $\hat{P}(t_0, t_0) = 1$, $\hat{P}(t+T, t_0) = \hat{P}(t, t_0)$, where $T = 2\pi/\omega$ is the period of the Hamiltonian. $\hat{G}(t_0)$ depends on the initial time t_0. Its eigenvalues are the quasienergies [262] of the system, which do not depend on t_0. The actual calculation of $\hat{G}(t_0)$ is in general complicated. For very large frequencies ω, the effective Hamiltonian calculated in the previous section is identical to one of the quasienergy operators $\hat{G}(t_1)$ with $t_0 \leq t_1 \leq t_0 + T$ [256, 257] and the quasienergies can be interpreted as mean energies.

5.3.2 The effect of a phase hop

We now address the question how a phase hop affects the mean motion of the quantum particle. Suppose that a phase hop of $\Delta\varphi$ is induced at the time $t_{\rm ph}$. As the effective Hamiltonian does not depend on the phase of V, the equation of motion (5.19) for the particle's mean-motion wave function remains unchanged. However, the mean-motion wave function itself changes due to the continuity of the particle's real wave function at $t = t_{\rm ph}$: for $t \neq t_{\rm ph}$ the latter is a product of a phase factor and the mean-motion wave function (Eq. (5.16)). As the phase factor (e^{-iF}) depends on the phase of V through

158   Chapter 5. Particle motion in rapidly oscillating potentials

F (Eq. (5.18)), it changes instantaneously and thus involves a corresponding change of the mean-motion wave function. In the following we calculate the mean-motion wave function ϕ_{new} for times after the phase hop and we derive the resulting changes of the mean-motion observables.

The condition of continuity for the particle's real wave function yields

$$\phi_{\text{new}}(x, t_{\text{ph}}) = e^{i\Delta F(x, \tau_{\text{ph}})} \phi(x, t_{\text{ph}}) \qquad (5.31)$$

(the notation $\lim_{t \to t_{\text{ph}}, t < t_{\text{ph}}} \phi(x,t) \equiv \phi(x, t_{\text{ph}})$ and $\lim_{t \to t_{\text{ph}}, t > t_{\text{ph}}} \phi_{\text{new}}(x,t) \equiv \phi_{\text{new}}(x, t_{\text{ph}})$ is used), where

$$\Delta F(x, \tau_{\text{ph}}) \equiv F(x, \tau_{\text{ph}} + \Delta\varphi) - F(x, \tau_{\text{ph}}) = \frac{1}{\hbar\omega} \int_0^{\tau_{\text{ph}}} [V_1(x, \tau + \Delta\varphi) - V_1(x, \tau)] \, d\tau. \quad (5.32)$$

Applying Eq. (5.19) leads to

$$\phi_{\text{new}}(x, t) = e^{-i(t-t_{\text{ph}})\hat{H}_{\text{eff}}(x)/\hbar} e^{i\Delta F(x, \tau_{\text{ph}})} \phi(x, t_{\text{ph}}). \qquad (5.33)$$

Combined with Eq. (5.22), Eq. (5.33) allows the entire description of the phase hop, as the particle's mean-motion wave function is known for all times.

The effect of the phase hop on ϕ involves an instantaneous change of some mean-motion observables, which, for a given mean-motion observable $\hat{O}$, is given by $\Delta \overline{\langle O \rangle} = \langle \phi_{\text{new}}(x, t_{\text{ph}}) | \hat{O} | \phi_{\text{new}}(x, t_{\text{ph}}) \rangle - \langle \phi(x, t_{\text{ph}}) | \hat{O} | \phi(x, t_{\text{ph}}) \rangle$. Using Eqs. (5.22) and (5.31) we find

$$\begin{aligned}
\Delta \overline{\langle x \rangle} &= 0, & (5.34) \\
\Delta \overline{\langle p \rangle} &= -i\hbar \langle \phi(x, t_{\text{ph}}) | \left[e^{-i\Delta F(x, \tau_{\text{ph}})} \frac{\partial}{\partial x} e^{i\Delta F(x, \tau_{\text{ph}})} - \frac{\partial}{\partial x} \right] | \phi(x, t_{\text{ph}}) \rangle \\
&= \hbar \int_{-\infty}^{\infty} |\phi(x, t_{\text{ph}})|^2 \Delta F'(x, \tau_{\text{ph}}) \, dx. & (5.35)
\end{aligned}$$

Thus, as for the classical case (see Eqs. 5.10 and 5.11), the effect of a phase hop can be visualized as being the result of an imaginary collision. The particle's mean-energy,

5.3. Quantum motion in a rapidly oscillating potential

which is conserved before and after the phase hop, changes by

$$\begin{aligned}
\Delta \overline{\langle E \rangle} &= \langle \phi_{\text{new}}(x, t_{\text{ph}}) | \hat{H}_{\text{eff}}(x) | \phi_{\text{new}}(x, t_{\text{ph}}) \rangle - \langle \phi(x, t_{\text{ph}}) | \hat{H}_{\text{eff}}(x) | \phi(x, t_{\text{ph}}) \rangle \\
&= \langle \phi(x, t_{\text{ph}}) | \left[e^{-i \Delta F(x, \tau_{\text{ph}})} \hat{H}_{\text{eff}}(x) e^{i \Delta F(x, \tau_{\text{ph}})} - \hat{H}_{\text{eff}}(x) \right] | \phi(x, t_{\text{ph}}) \rangle \\
&= \frac{\hbar^2}{2m} \langle \phi(x, t_{\text{ph}}) | \left[-i \Delta F''(x, \tau_{\text{ph}}) - 2i \Delta F'(x, \tau_{\text{ph}}) \frac{\partial}{\partial x} + (\Delta F'(x, \tau_{\text{ph}}))^2 \right] | \phi(x, t_{\text{ph}}) \rangle \\
&= -\frac{\hbar^2}{2m} i \int_{-\infty}^{\infty} \phi(x, t_{\text{ph}})^* \phi(x, t_{\text{ph}}) \Delta F''(x, \tau_{\text{ph}}) \, dx \\
&\quad -\frac{\hbar^2}{m} i \int_{-\infty}^{\infty} \phi(x, t_{\text{ph}})^* \phi'(x, t_{\text{ph}}) \Delta F'(x, \tau_{\text{ph}}) \, dx \\
&\quad +\frac{\hbar^2}{2m} i \int_{-\infty}^{\infty} \phi(x, t_{\text{ph}})^* \phi(x, t_{\text{ph}}) (\Delta F'(x, \tau_{\text{ph}}))^2 \, dx \\
&= \frac{\hbar^2}{2m} i \int_{-\infty}^{\infty} [\phi'(x, t_{\text{ph}})^* \phi(x, t_{\text{ph}}) + \phi(x, t_{\text{ph}})^* \phi'(x, t_{\text{ph}})] \Delta F'(x, \tau_{\text{ph}}) \, dx \\
&\quad -\frac{\hbar^2}{m} i \int_{-\infty}^{\infty} \phi(x, t_{\text{ph}})^* \phi'(x, t_{\text{ph}}) \Delta F'(x, \tau_{\text{ph}}) \, dx \\
&\quad +\frac{\hbar^2}{2m} i \int_{-\infty}^{\infty} |\phi(x, t_{\text{ph}})|^2 (\Delta F'(x, \tau_{\text{ph}}))^2 \, dx \\
&= \frac{\hbar^2}{2m} i \int_{-\infty}^{\infty} [\phi'(x, t_{\text{ph}})^* \phi(x, t_{\text{ph}}) - \phi(x, t_{\text{ph}})^* \phi'(x, t_{\text{ph}})] \Delta F'(x, \tau_{\text{ph}}) \, dx \\
&\quad +\frac{\hbar^2}{2m} \int_{-\infty}^{\infty} |\phi(x, t_{\text{ph}})|^2 (\Delta F'(x, \tau_{\text{ph}}))^2 \, dx
\end{aligned} \tag{5.36}$$

(the notation $\lim_{t \to t_{\text{ph}}, t < t_{\text{ph}}} \phi'(x, t) \equiv \phi'(x, t_{\text{ph}})$ is used), where we performed integration by parts, assuming that the particle is trapped, which implies $\lim_{x \to \pm \infty} |\phi(x, t_{\text{ph}})|^2 = 0$. Inspection of the rhs of Eq. (5.36) shows that $\Delta \overline{\langle E \rangle}$ is always non-negative if the particle is in a stationary mean-motion state ϕ_n (i.e., its mean motion is in an eigenstate ϕ_n of $\hat{H}_{\text{eff}}$):

$$\Delta \overline{\langle E_n \rangle} \geq 0 \, , \tag{5.37}$$

since it is $\phi_n'^* \phi_n = \phi_n^* \phi_n'$ [6]. Using the picture of the imaginary collision, Eq. (5.37) also directly follows from the fact that stationary mean-motion states have $\overline{\langle p \rangle} = 0$. For the Paul trap potential (5.2) the fact that the energy change $\Delta \overline{\langle E_n \rangle}$ can be non-zero (and even be very large as is shown in the subsequent paragraph) marks a difference to the classical regime, since for a classical particle whose mean motion is at rest, $\Delta \overline{E}_{\text{class.}}$ is always zero [124]. This difference is a direct consequence of Heisenberg's uncertainty

---

[6] A stationary mean-motion state obeys the relation $\phi_n'(x, t_{\text{ph}})^* \phi_n(x, t_{\text{ph}}) = \phi_n(x, t_{\text{ph}})^* \phi_n'(x, t_{\text{ph}})$, because as an eigenstate of a time-independent Hamiltonian it can be written in the form $\phi_n(x, t) = \xi_n(x) \exp(-iE_n t / \hbar)$. Thus, one has $\phi_n'(x, t) = \xi_n'(x) \exp(-iE_n t / \hbar)$ and thus $\phi_n'(x, t_{\text{ph}})^* \phi_n(x, t_{\text{ph}}) = \xi_n'(x)^* \xi_n(x) = \phi_n(x, t_{\text{ph}})^* \phi_n'(x, t_{\text{ph}})$.

principle which implies that in the quantum regime the particle's mean-position is spread around zero, leading—contrarily to the classical regime—to a non-vanishing micromotion (since $e^{-iF} \not\equiv 1$ for $x \neq 0$), thus giving rise to an effect of the phase hop. If the quantum particle is not in a stationary mean-motion state its mean-energy can, due to Eq. (5.36), be both increased and decreased:

$$\Delta \overline{\langle E \rangle} \gtreqless 0. \tag{5.38}$$

This case we will consider in more detail in Sec. 5.4.

Experimentally relevant are particles that are in stationary mean-motion states ϕ_n with mean-energy E_n. We now calculate how these states are affected by a phase hop. First we consider the model potentials. For the Paul-trap (5.2) we find

$$\Delta \overline{\langle E_n^{\mathrm{PT}} \rangle} = [\sin(\tau_{\mathrm{ph}} + \Delta\varphi) - \sin(\tau_{\mathrm{ph}})]^2 \, E_n^{\mathrm{PT}}, \tag{5.39}$$

where $E_n^{\mathrm{PT}} = \hbar\Omega(n+1/2)$ (Eq. (5.27)). Thus, the relative change $\Delta \overline{\langle E_n^{\mathrm{PT}} \rangle}/E_n^{\mathrm{PT}}$ is independent of n and of ω, and it can take values between 0 and 4. This demonstrates that in the Paul-trap a phase hop can have a strong effect on the particle's mean motion, even for arbitrarily large ω. For the rapidly scanning optical tweezers (5.3) we find accordingly

$$\Delta \overline{\langle E_n^{\mathrm{OT}} \rangle} = \frac{\omega_{\mathrm{ref}}^2}{\omega^2} [\sin(\tau_{\mathrm{ph}} + \Delta\varphi) - \sin(\tau_{\mathrm{ph}})]^2 \, E_0^{\mathrm{OT}}, \tag{5.40}$$

with $E_n^{\mathrm{OT}} = \hbar\omega_{\mathrm{osc}}(n+1/2)$ (Eq. (5.29)) and the reference frequency $\omega_{\mathrm{ref}} \equiv \sqrt{m\omega_{\mathrm{osc}}^3 x_0^2/\hbar}$. Note that the absolute change $\Delta \overline{\langle E_n^{\mathrm{OT}} \rangle}$ of the particle's mean-energy is independent of n and thus independent of the particle's mean-energy itself. The relative change $\Delta \overline{\langle E_n^{\mathrm{OT}} \rangle}/E_n^{\mathrm{OT}}$ can take values between 0 and $[4/(2n+1)]\,\omega_{\mathrm{ref}}^2/\omega^2$ and thus becomes negligible for $\omega \to \infty$. However, $\Delta \overline{\langle E_n^{\mathrm{OT}} \rangle}/E_n^{\mathrm{OT}}$ can still be large even if ω is as large as required by the validity condition of the underlying effective theory (i.e. for $\omega \gg \omega_{\mathrm{osc}}$). For example, for $\omega = \omega_{\mathrm{ref}}$ the energy change $\Delta \overline{\langle E_0^{\mathrm{OT}} \rangle}$ for the particle in the mean-motion ground state is identical to the corresponding change $\Delta \overline{\langle E_0^{\mathrm{PT}} \rangle}$ for the particle in the Paul-trap, provided that $\omega_{\mathrm{ref}} \gg \omega_{\mathrm{osc}}$. But this requirement can always be fulfilled with an adequate choice of parameters x_0 and ω_{osc} [7].

To generalize the above findings, consider first an arbitrary ROP with a vanishing

---

[7]For example, in a typical ultracold atom experiment [263], in which an optical tweezers is used to confine (lithium) atoms in the quantum regime, the experimental parameters are $m = 1.15 \cdot 10^{-26}$ kg, $\omega_{\mathrm{osc}} = 10^5$ Hz and the waist of the optical tweezers is $w_0 = 3 \cdot 10^{-5}$ m. If one rapidly modulated the position of such an optical tweezers with an amplitude of $x_0 = 3 \cdot 10^{-5}$ m, the value of the parameter ω_{ref} would be $\omega_{\mathrm{ref}} \approx 100\, \omega_{\mathrm{osc}}$.

5.3. Quantum motion in a rapidly oscillating potential

time-average. The mean potential energy of a particle in a stationary mean-motion state ϕ_n then is $E_n^{\text{pot}} = \langle\phi_n|V_{\text{eff}}|\phi_n\rangle$ with $V_{\text{eff}}(x) = \hbar^2\overline{F'(x,\tau)^2}/(2m)$ and the following relation holds:

$$\begin{aligned}
E_n^{\text{pot}} &< \max_{\tau_{\text{ph}}}\left[\langle\phi_n(x)|\tfrac{\hbar^2}{2m}F'(x,\tau_{\text{ph}})^2|\phi_n(x)\rangle\right] \\
&< \max_{\tau_{\text{ph}},\Delta\varphi}\left[\langle\phi_n(x)|\tfrac{\hbar^2}{2m}(\Delta F'(x,\tau_{\text{ph}}))^2|\phi_n(x)\rangle\right] \\
&= \max_{\tau_{\text{ph}},\Delta\varphi}\left[\Delta\overline{\langle E_n\rangle}\right].
\end{aligned} \quad (5.41)$$

Therefore a time τ_{ph} exists (within each period of V) for which a phase hop of a size $\Delta\varphi$ (with $0 \leq \Delta\varphi < 2\pi$) induces a change $\Delta\overline{\langle E_n\rangle}$ of the particle's mean-energy E_n which is greater than its mean potential energy E_n^{pot}. Since E_n^{pot} is in general a significant fraction of E_n, Eq. (5.41) shows that the phase hop can *always* be induced such that $\Delta\overline{\langle E_n\rangle}$ is large with respect to E_n, even for arbitrarily large ω.

In ROPs with a non-vanishing time-average a phase hop can only then lead to a significant change of the particle's mean-energy if ω is not too large, since for such ROPs the fraction of the particle's mean potential energy which equals the average kinetic energy stored in the particle's micromotion scales with ω^{-2}. This is expressed by Eq. (5.36), which yields that for stationary mean-motion states $\Delta\overline{\langle E_n\rangle}$ scales with ω^{-2} (see Eq. (5.40)).

We have seen that the phase hop can affect the particle's mean-motion wave function and observables. The mean-motion wave function was found to change in a non-continuous manner (Eq. (5.31)). At the same time we have seen that the effective Hamiltonian (5.20) of the mean-motion Schrödinger equation (5.19) is not affected by the phase hop, such that before and after the phase hop the system's stationary mean-motion states are identical. However, this does *not* mean that the particle does *not* change its state if it is in a stationary mean-motion state when a phase hop is induced. The phase hop can indeed induce transitions between stationary mean-motion states, $\phi_n \to \phi_m$, whose probabilities are given by

$$p_{n,m} \equiv |\langle\phi_{(n)}^{\text{new}}(x,t_{\text{ph}})|\phi_m(x)\rangle|^2 = \left|\int_{-\infty}^{\infty}\phi_n(x)\,\phi_m(x)\,e^{-i\Delta F(x,\tau_{\text{ph}})}\,dx\right|^2, \quad (5.42)$$

(implying $p_{n,m} = p_{m,n}$), where $\phi_{(n)}^{\text{new}}$ denotes the particle's mean-motion state after the phase hop. For the potentials (5.2) and (5.3), the $p_{n,m}$ can be calculated analytically, since the ϕ_n are explicitly known (see Eqs. (5.26) and (5.28)). For the Paul trap, Eq. (5.42) implies $p_{n,m}^{\text{PT}} = 0$, for $(n+m) =$ odd. This means that a phase hop induces only transitions between alternate levels, *i.e.* the selection rule is $n \to n \pm 2k$ (with

162 Chapter 5. Particle motion in rapidly oscillating potentials

integer k). Thus, it cannot mix the populations in even- and odd-numbered levels.

In particular, the $p_{0,m}$ are of experimental relevance as the mean-motion ground state ϕ_0 can be prepared with a high precision and can be easily probed. For the Paul-trap (5.2) we find

$$p_{0,m}^{\text{PT}} = \left(\frac{m!}{2^{\frac{3m}{2}} \left(\frac{m}{2}\right)! \left(\frac{m}{2}\right)!} \right) \frac{\delta^m}{\left(1 + \frac{\delta^2}{2}\right)^{\frac{m+1}{2}}}, \quad \text{for } m = \text{even}, \quad (5.43)$$

$$p_{0,m}^{\text{PT}} = 0, \quad \text{for } m = \text{odd}, \quad (5.44)$$

with $\delta = \sin(\tau_{\text{ph}} + \Delta\varphi) - \sin(\tau_{\text{ph}})$. In typical single ion experiments, the $p_{0,m}$ could be directly measured using resolved Raman sideband spectroscopy [234, 235]. Figure 5.4 shows that the probability for a particle to remain in the mean-motion ground state can be as small as 58%, demonstrating the significance of the effect of the phase hop. For the rapidly scanning optical tweezers (5.3) we find

$$p_{0,m}^{\text{OT}} = \frac{1}{2^m m!} \left(\frac{\omega_{\text{ref}}}{\omega} \delta \right)^{2m} e^{-\frac{1}{2}\left(\frac{\omega_{\text{ref}}}{\omega}\delta\right)^2}. \quad (5.45)$$

As $p_{0,0}^{\text{OT}} \to 1$ for $\omega \to \infty$, the effect of the phase hop becomes negligible for too large ω. However, Fig. 5.4 shows that for $\omega = \omega_{\text{ref}}$ the effect of the phase hop is still significant. For weakly interacting bosonic quantum gases, $p_{0,0}^{\text{OT}}$ could be measured using a Bose-Einstein condensate [251]. For degenerate atom gases, phase hops could offer a tool to more quickly increase the energy (see Eq. (5.40)) and thus, after thermalization, the temperature in a controlled way without having to change or to switch off-and-back-on the trapping potential and to in-between await an expansion of the gas [264].

Let us also briefly consider the effect of the phase hop on an excited stationary mean-motion state, e.g. on the state ϕ_2^{PT} in the Paul trap. The transition probabilities $p_{2,m}$ are

$$p_{2,m}^{\text{PT}} = \left(\frac{m!}{2^{\frac{3m+4}{2}} \left(\frac{m}{2}\right)! \left(\frac{m}{2}\right)!} \right) \frac{\delta^{m-2}(\delta^2 - 2m)^2}{\left(1 + \frac{\delta^2}{2}\right)^{\frac{m+3}{2}}}, \quad \text{for } m = \text{even}. \quad (5.46)$$

Thus, the transition probability $p_{2,0}^{\text{PT}}$ from the stationary mean-motion state $\phi_2^{\text{PT}}(x)$ to the mean-motion ground state $\phi_0^{\text{PT}}(x)$, which has a lower mean-energy, can be non-zero. This is not in contradiction to the previously found result that a particle in a stationary mean-motion state can only increase (the expectation value of) its energy when a phase hop is induced (see Eq. (5.37)). It is thus indeed possible that the phase hop dissipates energy from a particle in a stationary mean-motion state. Another conclusion can be drawn from Eq. (5.46). $p_{2,2}^{\text{PT}}$ vanishes for $\delta = \pm 2$, which demonstrates that, for correctly

5.3. Quantum motion in a rapidly oscillating potential

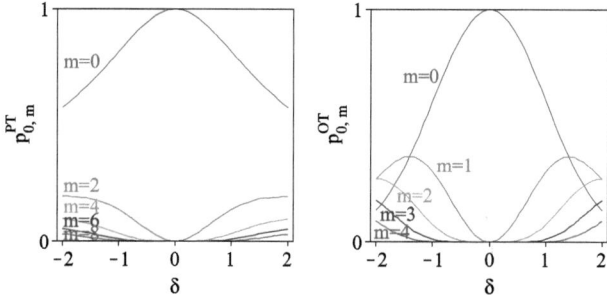

Figure 5.4: The experimentally measurable [234, 235] probabilities for the transitions between the mean-motion ground state ϕ_0 and the mean-motion states ϕ_m induced by a phase hop for a particle in the Paul-trap (5.43) (left) and the rapidly scanning optical tweezers (5.45) (right, with $\omega = \omega_{\text{ref}}$) as a function of the parameter $\delta = \sin(\tau_{\text{ph}} + \Delta\varphi) - \sin(\tau_{\text{ph}})$.

chosen parameters τ_{ph} and $\Delta\varphi$, the phase hop can cause the particle to leave its state $\phi_2^{\text{PT}}(x)$ with a probability of 100%, and thus that the phase hop can cause events to occur with an absolute certainty.

The manipulation by phase hops can be made more effective by inducing several phase hops successively. Figure 5.5 shows the transition probability $p_{0,0}^{\text{PT}}$ for two successively induced phase hops of size $\Delta\varphi = \pi$ as a function of their time delay. Although the effects of the two phase hops on the particle's (2π-periodic) Hamiltonian (5.14) cancel each other, their effects on the particle's mean motion do not necessarily cancel and can even be more significant as in the case of a single phase hop. As a matter of fact, it is in general not even true that the effect of the first phase hop can be canceled by the second. This holds only for ROPs with a harmonic effective potential. The revivals of $p_{0,0}^{\text{PT}}$ in Fig. 5.5 show that for the Paul trap the effect of a phase hop is reversible.

5.3.3 Numerical simulations

To countercheck our analytical predictions, we performed numerical simulations for a particle in the Paul-trap potential (5.2) by integrating the full time-dependent Schrödinger equation (5.13) using the Shampine-Gordon routine. Figure 5.6 shows the time-evolution of the experimentally measurable root-mean-square (rms)-deviation $\Delta x \equiv \sqrt{\langle\psi|\hat{x}^2|\psi\rangle - \langle\psi|\hat{x}|\psi\rangle^2}$ of the particle's position when influenced by a phase hop. Δx is given in units of its theoretically predicted mean-value before the phase hop is

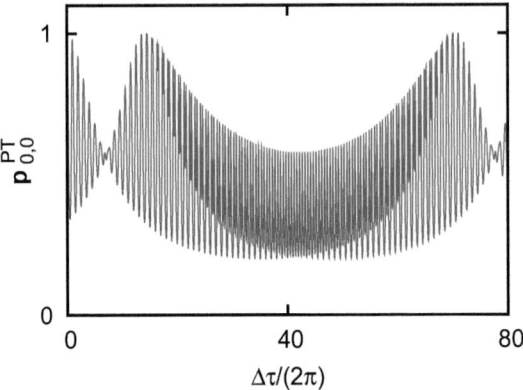

Figure 5.5: The experimentally measurable [234, 235] probability for a particle in the Paul-trap (5.2) to return to the mean-motion ground state ϕ_0 when two phase hops of size $\Delta\varphi = \pi$ are successively induced (at times $\tau_{\text{ph}}^{(1)} = \pi/2$ and $\tau_{\text{ph}}^{(2)} = \tau_{\text{ph}}^{(1)} + \Delta\tau$) as a function of $\Delta\tau$. Although one might naively expect those two phase-hops cancel each other, the second phase hop can, in fact, further reduce the probability to return to the ground state. Even parameters for which $p_{0,0}^{\text{PT}} = 1$ are interesting experimentally: they can be used to verify if the system indeed is described by dissipation-less quantum mechanics.

induced, which is given by

$$\Delta x_{0,\text{eff}} \equiv \sqrt{\langle \phi_0^{\text{PT}}(x)|\hat{x}^2|\phi_0^{\text{PT}}(x)\rangle - \langle \phi_0^{\text{PT}}(x)|\hat{x}|\phi_0^{\text{PT}}(x)\rangle^2} = \sqrt{\frac{1}{2}\frac{\hbar}{m\Omega}}. \quad (5.47)$$

The initial state was chosen to be $\psi(x,0) = e^{-iF^{\text{PT}}(x,0)}\phi_0^{\text{PT}}(x)$, which determines the particle to be in the mean-motion ground state ϕ_0^{PT}. Figure 5.6 shows very good agreement between numerics and the theoretical mean-motion predictions derived via computer-algebra from Eqs. (5.22) and (5.33). In particular, Fig. 5.6 confirms that a phase hop can have a strong effect on the particle's motion, since it shows that the time-evolution of Δx strongly depends on the phase hop size $\Delta\varphi$. As well, Fig. 5.6 shows that the particle's motion possesses two time scales—the time scale that is defined by the small oscillations of Δx around its mean (which represent the micromotions) and the time scale defined by the oscillations of its mean. As can be seen in the figure, the oscillations of the mean of Δx after the phase hop have a period of $2\pi/\Omega$, which agrees with what has been stated in Sec. 5.3.2 that the particle's mean motion is not only before, but also after the phase hop governed by a Schrödinger equation with the time-independent effective Hamiltonian.

The driving frequency used in the simulation is $\omega \approx 70\,\Omega$, and thus the numerics

5.3. Quantum motion in a rapidly oscillating potential

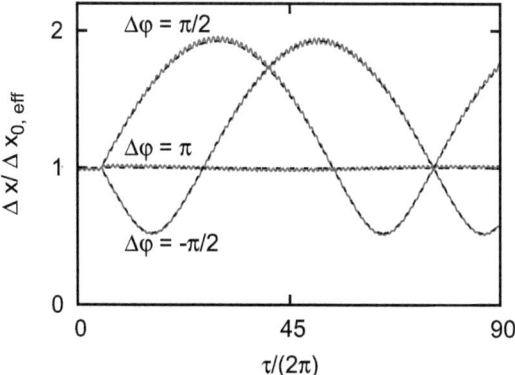

Figure 5.6: Time-evolution of the rms-deviation Δx for a particle in the Paul-trap (5.2) which initially is in the mean-motion ground state, affected by a phase hop of different sizes $\Delta\varphi$ (normalized to the theoretically predicted initial mean-motion value $\Delta x_{0,\text{eff}} = \sqrt{\hbar/(2m\Omega)}$). Solid (red) curves: results of numerical simulations, dashed (black) curves: theoretical mean-motion prediction. Driving frequency: $\omega \approx 70\,\Omega$.

demonstrate that the theoretical predictions for the limit $\omega \to \infty$ even hold for finite ω. For larger ω the agreement between both approaches would even be better. Further, the numerics allowed us to obtain a practical definition for "instantaneous": in experiments the phase-hop must happen on time-scales smaller than the period of V. As has been demonstrated in the recent experiment in Amsterdam [126], it is indeed possible to generate such small time-scales, although the period of the ROP is already very small.

In Sec. 5.3.2 we have explicitly calculated the probability for a particle that initially is in the mean-motion ground state $\phi_0^{\text{PT}}(x)$ to remain in this state when a phase hop is induced (see Eq. (5.43)). We obtained that for the phase hop parameter $\delta = \pm 2$ this probability equals 58%. We now countercheck this prediction by a numerical simulation. The initial state is again chosen to be $\psi(x,0) = e^{-iF^{\text{PT}}(x,0)}\phi_0^{\text{PT}}(x)$ and a phase hop of the size $\Delta\varphi = \pi$ is induced at the time $\tau_{\text{ph}} = 10.5\,\pi$ (corresponding to $\delta = -2$). The time-evolution of the above-mentioned probability is then obtained from the numerically calculated $\psi(x,t)$ and $\psi^{\text{new}}(x,t)$ via

$$p_{0,0}^{\text{num}}(\tau) = \begin{cases} \left|\langle e^{iF^{\text{PT}}(x,\tau)}\psi(x,\tau/\omega)\mid \phi_0^{\text{PT}}(x)\rangle\right|^2 & \text{for } \tau < \tau_{\text{ph}}, \\ \left|\langle e^{iF_{\text{new}}^{\text{PT}}(x,\tau)}\psi^{\text{new}}(x,\tau/\omega)\mid \phi_0^{\text{PT}}(x)\rangle\right|^2 & \text{for } \tau > \tau_{\text{ph}}. \end{cases} \quad (5.48)$$

It is depicted in Fig. 5.7 (dashed (blue) curve) for time-intervals of a different length (left and right). One can see that it approximately changes its value from 1 to ~ 0.58, at the time when the phase hop is induced. This agrees with the corresponding theoretical predition, which is according to Eq. (5.43) given by

$$p_{0,0}^{\text{theor}}(\tau) = \begin{cases} 1 & \text{for } \tau < \tau_{\text{ph}}, \\ 0.58 & \text{for } \tau > \tau_{\text{ph}}. \end{cases} \qquad (5.49)$$

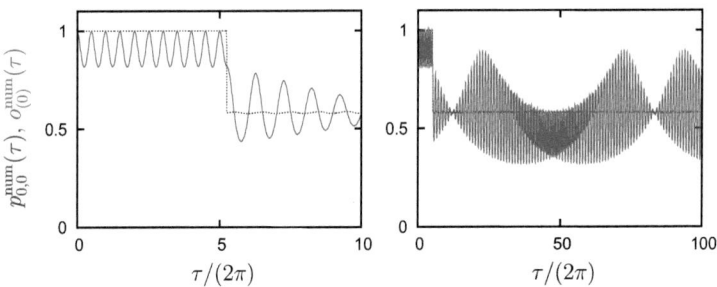

Figure 5.7: Dashed (blue) curve: time-evolution of the probability $p_{0,0}^{\text{num}}(\tau)$ (5.48) for a particle which initially is in the mean-motion ground state $\phi_0^{\text{PT}}(x)$ of the Paul-trap to remain in the mean-motion ground state under the influence of a phase hop of the size $\Delta\varphi = \pi$, induced at the time $\tau_{\text{ph}} = 10.5\pi$, obtained from a numerical simulation. Solid (red) curve: time-evolution of the modulus squared overlap $o_{(0)}^{\text{num}}(\tau)$ (5.50) of the particle's real state and its corresponding mean-motion state, obtained from a numerical simulation. The left and the right graph show the time-evolutions for time intervals of a different length. The potential's driving frequency was chosen to be $\omega \approx 70\,\Omega$.

In Sec. 5.3.2 we have shown that for the case of rapidly oscillating potentials with a vanishing mean the real wave function $\psi(x,t)$ of a trapped particle in general differs appreciably from its corresponding mean-motion wave function $\phi(x,t) = e^{iF(x,\tau)}\psi(x,t)$ for arbitrarily large driving frequencies ω, although $F(x,\tau)$ tends to zero as ω becomes large. To verify this conclusion, we calculate the time-evolution of the modulus squared overlap of the numerically calculated wave function $\psi(x,t)$ with its corresponding mean-motion wave function, which is given by $\phi_0^{\text{PT}}(x)$ before the phase hop and by $\phi_{(0)}^{\text{PT,new}}(x,t)$ after the phase hop. The modulus squared overlap is thus given by

$$o_{(0)}^{\text{num}}(\tau) = \begin{cases} \left|\langle \psi(x,\tau/\omega) \mid \phi_0^{\text{PT}}(x) \rangle\right|^2 & \text{for } \tau < \tau_{\text{ph}}, \\ \left|\langle \psi^{\text{new}}(x,\tau/\omega) \mid \phi_{(0)}^{\text{PT,new}}(x,\tau/\omega) \rangle\right|^2 & \text{for } \tau > \tau_{\text{ph}}, \end{cases} \qquad (5.50)$$

and shown in Fig. 5.7 (solid (red) curve) for time-intervals of a different length (left and right). One can see that $o_{(0)}^{\text{num}}(\tau)$ is periodic before the phase hop with the same period as $(F^{\mathrm{PT}}(x,\tau))^2$, indicating that the corresponding mean-motion state of $\psi(x,t)$ is the mean-motion ground state $\phi_0^{\mathrm{PT}}(x)$. After the phase hop, this periodicity is no longer given, indicating that the corresponding mean-motion state of $\psi^{\mathrm{new}}(x,t)$ after the phase hop, $\phi_{(0)}^{\mathrm{PT,new}}(x,t)$, is not a stationary mean-motion eigenstate. After the phase hop, $o_{(0)}^{\mathrm{num}}(\tau)$ is only periodic with the much larger period $2\pi/\Omega$, which agrees with what has been stated in Sec. 5.3.2 that the particle's mean motion is not only before, but also after the phase hop governed by a Schrödinger equation with the time-independent effective Hamiltonian. In Fig. 5.7 one can further see that $o_{(0)}^{\mathrm{num}}(\tau)$ performs large oscillations. These large oscillations are *not* the consequence of a too small chosen value for the driving frequency ω, since in Fig. 5.7 the same ω was used as in Fig. 5.6, whereas Fig. 5.6 proved that ω was chosen large enough to validate the underlying theory. In fact, the phenomenon that a phase hop can have a strong effect on the particle's motion, just originates from these large oscillations of $o_{(0)}^{\mathrm{num}}(\tau)$ for $\tau < \tau_{\mathrm{ph}}$.

5.4 Consistency between classical and quantum mechanical results

The results presented in Sec. 5.3 must be consistent with the classical results presented in Sec. 5.2. To countercheck this we calculate $\Delta\overline{\langle E \rangle}$ for the case that the mean motion of a quantum particle in the model potentials of the Paul-trap and the oscillating tweezers potential is in a *coherent* state [234, 235]. We then show that the classical limit of $\Delta\overline{\langle E \rangle}$ equals the corresponding classical result ΔE.

5.4.1 Coherent states

A quantum state of a particle in the harmonic oscillator potential

$$V^{\mathrm{HO}}(x) = \frac{1}{2} m\Omega^2 x^2 \qquad (5.51)$$

is called a coherent state, ϕ_{CS}, if it obeys

$$\begin{aligned}
\Delta x &\equiv \sqrt{\langle \phi_{\mathrm{CS}}|\hat{x}^2|\phi_{\mathrm{CS}}\rangle - \langle \phi_{\mathrm{CS}}|\hat{x}|\phi_{\mathrm{CS}}\rangle^2} = \sqrt{\frac{\hbar}{2m\Omega}}, \\
\Delta p &\equiv \sqrt{\langle \phi_{\mathrm{CS}}|\hat{p}^2|\phi_{\mathrm{CS}}\rangle - \langle \phi_{\mathrm{CS}}|\hat{p}|\phi_{\mathrm{CS}}\rangle^2} = \sqrt{\frac{\hbar m\Omega}{2}},
\end{aligned} \qquad (5.52)$$

where $\hat{x} = x$ and $\hat{p} = -i\hbar\partial/\partial x$ denote the position and momentum operators, respectively [265, 266]. Thus a coherent state is a quantum state which minimizes Heisenberg's uncertainty relation $\Delta x\,\Delta p \geq \hbar/2$. A coherent state can be written in the explicit form [265, 266]

$$\phi_{\text{CS}}(x) = \left(\frac{m\Omega}{\pi\hbar}\right)^{\frac{1}{4}} e^{\frac{i}{\hbar}\langle p\rangle x} e^{-\frac{m\Omega}{2\hbar}(x-\langle x\rangle)^2}, \qquad (5.53)$$

with $\langle\phi_{\text{CS}}|\hat{x}|\phi_{\text{CS}}\rangle = \langle x\rangle$ and $\langle\phi_{\text{CS}}|\hat{p}|\phi_{\text{CS}}\rangle = \langle p\rangle$. It describes a quantum particle whose position and momentum have the expectation values $\langle x\rangle$ and $\langle p\rangle$, respectively. For a coherent state, these expectation values evolve in time according to the laws of classical mechanics and the minimum uncertainty property $\Delta x\,\Delta p = \hbar/2$ is conserved. Thus a coherent state yields the most precise quantum mechanical description of the motion of a classical particle in the harmonic oscillator potential (5.51). An example for a coherent state is the ground state

$$\phi_0(x) = \left(\frac{m\Omega}{\pi\hbar}\right)^{\frac{1}{4}} e^{-\frac{m\Omega}{2\hbar}x^2}, \qquad (5.54)$$

for which $\langle x\rangle = \langle p\rangle = 0$.

5.4.2 Effect of phase hop on a coherent mean-motion state

The mean motion of a quantum particle in both, the model potential (5.2) and (5.3), is governed by a time-independent *harmonic* effective potential. Therefore, the mean motion of a particle in these potentials possesses coherent states.

For the Paul-trap potential (5.2) the coherent mean-motion states are given by (see Eq. (5.53))

$$\phi_{\text{CS}}^{\text{PT}}(x) = \left(\frac{m\Omega}{\pi\hbar}\right)^{\frac{1}{4}} e^{\frac{i}{\hbar}\overline{\langle p\rangle}x} e^{-\frac{m\Omega}{2\hbar}(x-\overline{\langle x\rangle})^2}, \qquad (5.55)$$

where Eq. (5.55) describes a quantum particle with mean-position $\overline{\langle x\rangle}$, mean-momentum $\overline{\langle p\rangle}$ and mean-energy

$$\overline{\langle E_{\text{CS}}^{\text{PT}}\rangle} = \frac{\langle\phi_{\text{CS}}|\hat{p}^2|\phi_{\text{CS}}\rangle}{2m} + \langle\phi_{\text{CS}}|V_{\text{eff}}^{\text{PT}}(\hat{x})|\phi_{\text{CS}}\rangle = \frac{\overline{\langle p\rangle}^2}{2m} + V_{\text{eff}}^{\text{PT}}(\overline{\langle x\rangle}) + \frac{1}{2}\hbar\Omega. \qquad (5.56)$$

When a phase hop is induced the coherent state changes to (see Eq. (5.31))

$$\phi_{(\text{CS})}^{\text{PT,new}}(x) = e^{\frac{im\Omega\delta}{\hbar\sqrt{2}}x^2} \phi_{\text{CS}}^{\text{PT}}(x), \qquad (5.57)$$

5.4. Consistency between classical and quantum mechanical results

with $\delta = \sin(\tau_{\text{ph}} + \Delta\varphi) - \sin(\tau_{\text{ph}})$ and the particle's mean-energy changes by (see Eq. (5.36))

$$\begin{aligned}
\Delta \overline{\langle E_{\text{CS}}^{\text{PT}} \rangle} &= \frac{\hbar^2}{2m} i \int_{-\infty}^{\infty} \phi_{\text{CS}}^{\text{PT}\,\prime}(x)^* \phi_{\text{CS}}^{\text{PT}}(x) \left(\frac{\sqrt{2}m\Omega\delta}{\hbar} x \right) dx \\
&\quad - \frac{\hbar^2}{2m} i \int_{-\infty}^{\infty} \phi_{\text{CS}}^{\text{PT}}(x)^* \phi_{\text{CS}}^{\text{PT}\,\prime}(x) \left(\frac{\sqrt{2}m\Omega\delta}{\hbar} x \right) dx \\
&\quad + \frac{\hbar^2}{2m} \int_{-\infty}^{\infty} |\phi_{\text{CS}}^{\text{PT}}(x)|^2 \left(\frac{\sqrt{2}m\Omega\delta}{\hbar} x \right)^2 dx, \\
&= \frac{i}{\sqrt{2}} \hbar\Omega\delta \int_{-\infty}^{\infty} \left(-\frac{i}{\hbar} \overline{\langle p \rangle} - \frac{m\Omega}{\hbar}(x - \overline{\langle x \rangle}) \right) \phi_{\text{CS}}^{\text{PT}}(x)^* \phi_{\text{CS}}^{\text{PT}}(x) x\, dx \\
&\quad - \frac{i}{\sqrt{2}} \hbar\Omega\delta \int_{-\infty}^{\infty} \phi_{\text{CS}}^{\text{PT}}(x)^* \left(\frac{i}{\hbar} \overline{\langle p \rangle} - \frac{m\Omega}{\hbar}(x - \overline{\langle x \rangle}) \right) \phi_{\text{CS}}^{\text{PT}}(x) x\, dx \\
&\quad + m\Omega^2 \delta^2 \int_{-\infty}^{\infty} |\phi_{\text{CS}}^{\text{PT}}(x)|^2 x^2\, dx, \\
&= \sqrt{2}\, \Omega \overline{\langle x \rangle}\, \overline{\langle p \rangle} \delta + m\Omega^2 \langle \phi_{\text{CS}}^{\text{PT}} | \hat{x}^2 | \phi_{\text{CS}}^{\text{PT}} \rangle \delta^2. \quad (5.58)
\end{aligned}$$

In comparison, a classical particle that has the mean-position X and the mean-momentum $P = m\dot{X}$ would, according to Eq. (5.12), change its mean-energy by

$$\Delta E^{\text{PT}} = \sqrt{2}\, \Omega X P \delta + m\Omega^2 X^2 \delta^2. \quad (5.59)$$

Equation (5.59) is the classical limit of Eq. (5.58), since it is

$$\lim_{\hbar \to 0} \overline{\langle x \rangle} = X, \quad (5.60)$$

$$\lim_{\hbar \to 0} \overline{\langle p \rangle} = P, \quad (5.61)$$

$$\lim_{\hbar \to 0} \langle \phi_{\text{CS}}^{\text{PT}} | \hat{x}^2 | \phi_{\text{CS}}^{\text{PT}} \rangle = \lim_{\hbar \to 0} \langle \phi_{\text{CS}}^{\text{PT}} | \hat{x} | \phi_{\text{CS}}^{\text{PT}} \rangle^2 = X^2, \quad (5.62)$$

$$\lim_{\hbar \to 0} \langle \phi_{\text{CS}}^{\text{PT}} | \hat{p}^2 | \phi_{\text{CS}}^{\text{PT}} \rangle = \lim_{\hbar \to 0} \langle \phi_{\text{CS}}^{\text{PT}} | \hat{p} | \phi_{\text{CS}}^{\text{PT}} \rangle^2 = P^2. \quad (5.63)$$

This demonstrates that the quantum mechanical results of Sec. 5.3 agree with the classical ones presented in Sec. 5.2. Note that a phase hop destroys a coherent state for the Paul-trap potential, since right after the phase hop it is

$$\Delta x_{\text{new}}^{\text{PT}} \equiv \sqrt{\langle \phi_{\text{(CS)}}^{\text{PT,new}} | \hat{x}^2 | \phi_{\text{(CS)}}^{\text{PT,new}} \rangle - \langle \phi_{\text{(CS)}}^{\text{PT,new}} | \hat{x} | \phi_{\text{(CS)}}^{\text{PT,new}} \rangle^2} = \sqrt{\frac{\hbar}{2m\Omega}},$$

$$\Delta p_{\text{new}}^{\text{PT}} \equiv \sqrt{\langle \phi_{\text{(CS)}}^{\text{PT,new}} | \hat{p}^2 | \phi_{\text{(CS)}}^{\text{PT,new}} \rangle - \langle \phi_{\text{(CS)}}^{\text{PT,new}} | \hat{p} | \phi_{\text{(CS)}}^{\text{PT,new}} \rangle^2} = \sqrt{\frac{\hbar m\Omega}{2} + \hbar m\Omega \delta^2}.$$

Furthermore, $\Delta x_{\text{new}}^{\text{PT}}$ and $\Delta p_{\text{new}}^{\text{PT}}$ are not conserved in time. This can be seen also from Eq. (5.39), as it implies that the phase hop affects differently the wave function components of a coherent state, thereby destroying their coherence.

For the oscillating tweezers potential (5.3) the coherent mean-motion states are given by (see Eq. (5.53))

$$\phi_{\text{CS}}^{\text{OT}}(x) = \left(\frac{m\omega_{\text{osc}}}{\pi\hbar}\right)^{\frac{1}{4}} e^{\frac{i}{\hbar}\overline{\langle p\rangle}x} e^{-(x-\overline{\langle x\rangle})^2 m\omega_{\text{osc}}/(2\hbar)}. \tag{5.64}$$

When a phase hop is induced the coherent state changes to

$$\phi_{(\text{CS})}^{\text{OT,new}}(x) = e^{i[m\omega_{\text{osc}}^2 x_0 x \delta/(\hbar\omega)+\beta]} \phi_{\text{CS}}^{\text{OT}}(x), \tag{5.65}$$

with an unphysical constant $\beta = m\omega_{\text{osc}}^2 x_0^2 \left[\sin(2(\tau_{\text{ph}}+\Delta\varphi)) - \sin(2\tau_{\text{ph}})\right]/(8\hbar\omega)$. The particle's mean-energy in thus found to change by

$$\Delta\overline{\langle E_{\text{CS}}^{\text{OT}}\rangle} = \frac{\omega_{\text{osc}}^2 x_0 \overline{\langle p\rangle}\delta}{\omega} + \frac{m\omega_{\text{osc}}^4 x_0^2 \delta^2}{2\omega^2} \tag{5.66}$$

which, in the classical limit, goes over into the classical result $\Delta E^{\text{PT}} = \omega_{\text{osc}}^2 x_0 P\delta/\omega + m\omega_{\text{osc}}^4 x_0^2 \delta^2/(2\omega^2)$ derived from Eq. (5.58). Inspection of Eq. (5.65) shows that $\phi_{(\text{CS})}^{\text{OT,new}}$ equals $\phi_{\text{CS}}^{\text{OT}}$ of Eq. (5.64) when $\overline{\langle p\rangle}$ is replaced by $\overline{\langle p\rangle} + \Delta\overline{\langle p\rangle}$, with $\Delta\overline{\langle p\rangle} = m\omega_{\text{osc}}^2 x_0 \delta/\omega$ according to Eq. (5.35). Thus, a phase hop transforms a coherent state into a coherent state. For the oscillating tweezers potential, a phase hop thus conserves the coherence of coherent states. This conclusion can also be directly drawn from the fact that $\Delta\overline{\langle p\rangle}$ in Eq. (5.35) is independent of the state of the particle for the oscillating tweezers potential and thus affects in the same way the wave function components of a coherent state. This is consistent with the previously mentioned picture that a phase hop in the oscillating tweezers potential induces a Galilean transformation.

5.5 Conclusion

In this chapter we have presented a classical and quantum mechanical treatment of particle motion in a rapidly oscillating potentials under the influence of phase hops. We have computed the mean motion of a single particle for all times. For the quantum regime we have calculated the corresponding mean-motion wave function and the transition probabilities between stationary mean-motion states and we have shown that the particle's mean-energy can in general be both increased and decreased, except if it is in a stationary mean-motion state. Then the mean-energy can only be increased. In particular we have shown that the effect of a phase hop can be very strong.

5.5. Conclusion

Both for classical and for quantum particles, the induction of phase hops provides a powerful tool for particle manipulation. Besides its strong effect on the particle's mean motion, it has the following further appealing properties. First, its experimental implementation is very simple and does not require a change of an existing setup. Its feasibility has been demonstrated in the recent experiment in Amsterdam [126]. Second, its application is a controlled operation since a phase hop does not have an influence on the effective trapping potential, because that is independent of the phase of the ROP. Finally, it can be applied to any kind of trappable particles, because its mechanism is simply based on changes of the ROP and does thus not rely on particular internal properties of the particles as do laser-based manipulation methods. We intend to use the presented method in order to manipulate polar ^{6}Li^{40}K molecules.

We have demonstrated that in the Paul-trap potential a phase hop can dissipate mean energy of a particle by a factor, which is independent of its mean energy itself. Thus, it can dissipate more energy for particles with a higher energy. Phase hops might therefore offer a possibility to accelerate evaporative cooling for neutral particles. In the field of ultracold quantum gases, phase hops could also offer a convenient way to realize controlled heating.

The results presented in this chapter give an additional motivation for the use of ROPs in experiments. For example, one could take advantage of the described tool by inducing a rapid modulation on an existing static trap (modulating either the trap position or the trap depth). This would not appreciably change the particle's motion compared to the unmodulated case, but offer the possibility to significantly manipulate the trapped particles in a simple way.

In this chapter we have restricted our considerations to a single particle. Our main goal was to obtain analytical results, as these directly apply to a widely used and studied experimental system, the single ion in a Paul-trap [234, 235]. Due to the experimentally achievable pureness of this system, it can be used to precisely investigate phase hops and their possible applications experimentally. One of such possible applications could be to decelerate a single ion, which we here showed to be possible also in the quantum regime.

For future research, it will be interesting to study the effect of phase hops on ensembles of (interacting) particles. The extension of single particle effects in time-periodic systems [267, 268] to multi-particle systems is an active field of theoretical and experimental physics [269, 270]. An experimental investigation of the schemes proposed in this chapter has been done by the experiment in Amsterdam [126], where a Bose-Einstein condensate in a TOP trap was manipulated by inducing phase hops. It could be shown that the mean energy of the BEC could be decreased by a factor of 15 by inducing a single phase hop. Further investigations could be performed with cold ion clouds stored in a Paul-trap [235] or ultracold neutral atoms in rapidly scanning optical tweezers [250, 251] or in optical lattices subject to time-periodic perturbations [270, 271, 268, 272, 273]. In particular, in the latter system phase hops might offer a tool to manipulate and control the recently observed super-Bloch oscillations [273] as pointed out by Ref. [274]. Finally, phase hops could be further investigated using polar molecules trapped by electro- and magnetodynamic traps [236, 237].

Chapter 6

Conclusion

In this thesis a new experimental machine to study ultracold quantum gases composed of two different fermionic species is presented. It produces large atom number samples of ^{6}Li and ^{40}K atoms in an ultra-high vacuum environment which allows large optical access and strong magnetic confinement.

The study of fermionic mixtures composed of two different species is a new branch of ultracold Fermi gases. So far, only four other experimental groups are working with such systems [89, 43, 44, 42] (all using the mixture ^{6}Li-^{40}K), such that many of their aspects are still undiscovered. Furthermore, the mixture ^{6}Li-^{40}K promises rich physics due to the existence of both bosonic and fermionic stable isotopes for both species, offering the whole spectrum of possible quantum mixtures (boson/boson, fermion/boson, fermion/fermion). Once the mixture of our apparatus is brought to quantum degeneracy exciting experiments can be performed. The study of fermionic pairing in the mixture is of important interest as it might be relevant for high-T_c superconductors [275]. The unmatched Fermi surfaces resulting from the mass imbalance of the two species make symmetric BCS pairing impossible. New quantum phases with different pairing mechanisms are predicted, such as the previously mentioned Fulde-Ferrell-Larkin-Ovchinnikov (FFLO) state [78, 79] or the breached pair state [80, 81]. Other phenomena such as a crystalline phase transition [82] and the formation of long-lived trimers [83] are predicted. The large mass ratio between ^{6}Li and ^{40}K further allows the investigation of the influence of disorder on transport properties in optical lattices. The independent experimental control of the two species would make it possible to engineer a lattice which is deep enough to freeze the motion of the ^{40}K but not the ^{6}Li atoms. The ^{40}K atoms would act as random scatterers on the nearly free ^{6}Li atoms, providing a clean realization of the Anderson problem of localization by disorder [86], and its competition with interactions.

The presented machine has been designed to produce large atom number samples.

This goal was motivated by the fact that many experiments are usually limited by too small atom numbers, which result in stricter cooling requirements for the observation of quantum degeneracy and smaller signal-to-noise ratios, making quantitative studies difficult. We demonstrate in this thesis that the constructed machine is capable of producing the largest ^{6}Li-^{40}K dual-species MOT which has been reported in the literature. Two strategies have been applied in order to achieve this. First, the dual-species MOT is placed in an ultra-high vacuum environment, being continuously loaded from cold atomic beams, which originate from independent atom sources with a large flux. Second, the homo- and heteronuclear collisions have been minimized by using small magnetic field gradients and low light powers in the repumping light. Besides the large dual-species MOT, a magnetic transport has been implemented for a transfer of the atoms to an ultrahigh vacuum environment with large optical access, where evaporative cooling can be efficiently performed. So far the evaporative cooling procedure has not yet been realized, but is within close reach.

With the constructed apparatus we were able to create, for the first time, electronically excited weakly bound ^{6}Li^{40}K* molecules via photoassociation. The recorded photoassociation spectra yielded the binding energies of the 5-6 most weakly bound vibrational levels of the different molecular potentials. The observed resonances have been identified, which now allows the combination of our spectroscopic data with data obtained in conventional molecular spectroscopy experiments for deeply bound molecules. This will give access to more precise potential energy curves. In particular for the excited molecular potential $1(B)^1\Pi$, we believe that we have recorded all of the previously undetermined vibrational level positions, such that a complete set of vibrational level positions is now available for this potential. The gained information about the potential will allow us to find ways to efficiently transfer atoms to stable deeply bound ground-state molecules, based on the stimulated Raman adiabatic passage (STIRAP) technique [118, 85, 119]. The large formation rates for excited ^{6}Li^{40}K* molecules observed in our experiment and the recently reported efficient association of weakly bound ground-state ^{6}Li-^{40}K molecules via Feshbach resonances [93] promise high transfer efficiencies using this scheme. Once created, deeply bound ^{6}Li^{40}K molecules will, due to their strong long-range, anisotropic dipole-dipole interaction, give access to new quantum regimes. The effective coupling of the electric dipoles at moderate distances and the possibility to control the dipole moments by electric fields makes them good candidates for the realization of qubits for quantum computation [100]. They can further be used for fundamental tests like the measurement of the electron dipole moment [102, 103] or the study of time variation of fundamental constants [106, 276].

Besides the description of experimental work, this thesis contains a theoretical study of the motion of quantum particles in rapidly oscillating potentials. We have presented and studied a new method to manipulate particles in such potentials, which is based on inducing an instantaneous change of the potential's phase (phase hops). Phase hops offer the possibility to efficiently manipulate the particle's mean motion. Their experimental implementation is very simple and does not require a change of an existing setup. Its feasibility has been demonstrated in a recent experiment in Amsterdam [126]. The application of phase hops is a controlled operation since it does not affect the effective trapping potential, because that is independent of the phase of the rapidly oscillating potential. It can be applied to any kind of trappable particles, because its mechanism is simply based on changes of the rapidly oscillating potential and does thus not rely on particular internal properties of the particles as do laser-based manipulation methods. In particular, it could be applied to polar molecules stored in dynamic traps [236, 237]. Phase hops might also offer a possibility to accelerate evaporative cooling for neutral particles.

Appendix A

Determination of vapor pressure by light absorption

In the following we describe how the vapor pressure of potassium can be determined from the absorption profile of a resonant light beam. We have applied the presented method in order to determine the potassium vapor pressure in our 2D-MOT cell.

For low vapor pressures, for which the mean free path of a vapor atom is much larger than the dimensions of the vapor cell (typically below 10^{-4} mbar), the vapor pressure is determined by the temperature of the cell walls and is related to it according to the vapor pressure characteristics of the atom [162]. Since the vapor pressure grows exponentially with temperature, the determination of the vapor pressure via a temperature measurement of the cell walls would, however, be imprecise. In particular this method fails when parts of the cell walls have different temperatures. The pressure in a vapor cell can be more precisely determined from the transmission profile of a low intensity beam whose frequency is scanned through resonance.

For temperatures of the order of the room temperature the transmission profile through a vapor of potassium is a superposition of unresolved overlapping Doppler profiles, since potassium has a hyperfine structure, which is comparable to the Doppler width. For a correct analysis of the recorded transmission profile, the hyperfine structure thus has to be taken into account. It is, however, sufficient to consider only the hyperfine structure of the electronic ground states, since the hyperfine structure of the excited states is much smaller. Furthermore, the potassium vapor consists of several isotopes, with isotope shifts which are non-negligible with respect to the Doppler width, and thus their presence and abundances need to be taken into account also.

In the following we consider each K-isotope to be a three-level atom with one excited state and two ground states. The ground states are separated by the hyperfine ground state splittings of 461.7 MHz (^{39}K), 1285.8 MHz (^{40}K) and 254.1 MHz (^{41}K). The transition frequencies between the ground states and the excited state are referred to as $\nu_1^{(i)}$ and $\nu_2^{(i)}$, with $\nu_1^{(i)} = \nu_2^{(i)} + \Delta E_{\mathrm{HF}}^{(i)}$, where $i \in \{39, 40, 41\}$ denotes the corresponding K-isotope and $\Delta E_{\mathrm{HF}}^{(i)}$ the energy difference between the ground states. The transition frequencies are given by [277]

$$\nu_1^{(39)} = \nu_0 - 173.1\,\mathrm{MHz}, \tag{A.1}$$

$$\nu_2^{(39)} = \nu_0 + 288.6\,\mathrm{MHz}, \tag{A.2}$$

$$\nu_1^{(40)} = \nu_0 - 714.3\,\mathrm{MHz} + 125.58\,\mathrm{MHz}, \tag{A.3}$$

$$\nu_2^{(40)} = \nu_0 + 571.5\,\mathrm{MHz} + 125.58\,\mathrm{MHz}, \tag{A.4}$$

$$\nu_1^{(41)} = \nu_0 - 95.3\,\mathrm{MHz} + 235.0\,\mathrm{MHz}, \tag{A.5}$$

$$\nu_2^{(41)} = \nu_0 + 158.8\,\mathrm{MHz} + 235.0\,\mathrm{MHz}, \tag{A.6}$$

where $\nu_0 = 391.01$ THz denotes the frequency of the fine-structure transition $4S_{1/2} \to 4P_{3/2}$ of ^{39}K. The rightmost terms in Eqs.(A.3–A.6) account for the isotope shifts [277].

For a low beam intensity I_0 (*i.e.* $I_0 \ll I_{\mathrm{sat}}$) the relative absorption of the beam is independent of I_0. In this case the intensity of the beam as a function of the covered distance x inside the vapor cell is described by Lambert-Beer's law

$$I(x,\nu) = I_0 e^{-\left(0.895\frac{5}{8}\kappa_1^{(39)} + 0.895\frac{3}{8}\kappa_2^{(39)} + 0.04\frac{112}{387}\kappa_1^{(40)} + 0.04\frac{275}{387}\kappa_2^{(40)} + 0.065\frac{5}{8}\kappa_1^{(41)} + 0.065\frac{3}{8}\kappa_2^{(41)}\right)x}, \tag{A.7}$$

where the $\kappa_j^{(i)}$ depend on the beam frequency ν. The prefactors of the $\kappa_j^{(i)}$ correspond to the isotopic abundances in the sample (^{39}K: 89.5%, ^{40}K: 0.04%, ^{41}K: 6.5% in our 2D-MOT cell) and to the statistical weights of the transitions (derived from the Wigner-6j symbol [135, 278]). Assuming the vapor to be in thermal equilibrium at a temperature T^1, the $\kappa_j^{(i)}$ are given by

$$\kappa_1^{(i)}(\nu) = \kappa_0^{(i)} e^{-(\nu-\nu_1^{(i)})^2/(2\sigma_{(i)}^2)}, \tag{A.8}$$

$$\kappa_2^{(i)}(\nu) = \kappa_0^{(i)} e^{-(\nu-\nu_2^{(i)})^2/(2\sigma_{(i)}^2)}, \tag{A.9}$$

$$\kappa_0^{(i)}(\nu) = nh\nu \frac{\gamma^2}{4I_{\mathrm{sat}}\nu_0} \sqrt{\frac{m_{(i)}c^2}{2\pi k_B T}}, \tag{A.10}$$

$$\sigma_{(i)} = \nu_0 \sqrt{\frac{k_B T}{m_{(i)}c^2}}, \tag{A.11}$$

where $m_{(i)} = i \times 1.66 \times 10^{-27}$ kg is the mass of the isotope, n the atom density in

---
[1] When different parts of the vapor cell have different temperatures, this assumtion is not *a priori* justified. However, in our 2D-MOT cell we find that the recorded transmission profiles (see Fig. A.1) are well fitted by Eq. (A.7) and we conclude that the underlying assumption is valid in our case. The temperature T associated with each profile has to be considered an effective temperature.

the vapor, and $\Gamma = 2\pi \times 6.04\,\text{MHz}$, $h = 6.63 \times 10^{-34}\,\text{Js}$, $I_\text{sat} = 1.75\,\text{mW/cm}^2$, $k_\text{B} = 1.38 \times 10^{-23}\,\text{J/K}$ and $c = 3 \times 10^8\,\text{m/s}$. Fitting the recorded transmission profile with Eq. (A.7) yields n and T. The vapor pressure P is then obtained by the ideal gas equation $P = nkT$.

Figure A.1 shows different transmission profiles obtained in our 2D-MOT cell for different cell temperatures. For the measurement a peak intensity of $I_0 = 0.02 I_\text{sat}$ is used and the path of the probe beam inside the 2D-MOT cell has a length of 10.5 cm. The frequency of the probe beam is calibrated by an additional saturated absorption spectroscopy experiment. The blue curve in Fig. A.1 (labaled with (f)) corresponds to a vapor pressure of $2.3 \times 10^{-7}\,\text{mbar}$, which is the pressure which maximizes the flux of the 2D-MOT. It is obtained for an effective temperature of $\sim 45\,°\text{C}$. In the experimental setup, the used probe beam is permanently installed, allowing us to monitor the pressure in the cell at any time.

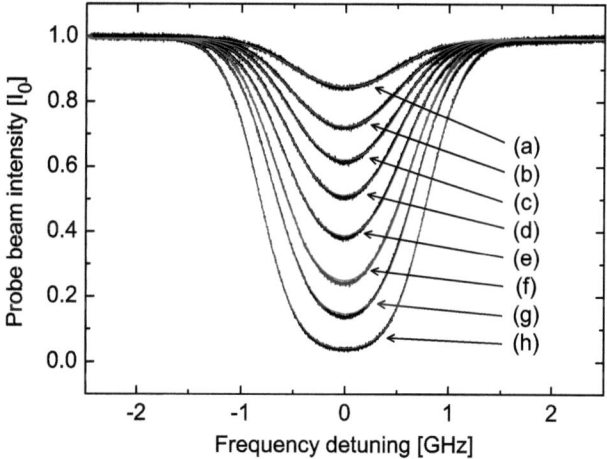

Figure A.1: Transmission profiles of a weak probe beam, which passes through the 2D-MOT cell with a path length of 10.5 cm for different cell temperatures. The recorded curves (black and blue) are fitted using Eq. (A.7) (smooth red curves) in order to determine the corresponding vapor pressure in the cell. In the graph, the frequency detuning refers to the frequency $\nu_\text{min} = \nu_0 - 0.09\,\text{GHz}$ for which the transmission is minimum (ν_0 denoting the frequency of the fine-structure transition $4S_{1/2} \rightarrow 4P_{3/2}$ of ^{39}K.). The vapor pressures corresponding to the curves are: (a) $3 \times 10^{-8}\,\text{mbar}$, (b) $6 \times 10^{-8}\,\text{mbar}$, (c) $9 \times 10^{-8}\,\text{mbar}$, (d) $1.3 \times 10^{-7}\,\text{mbar}$, (e) $1.8 \times 10^{-7}\,\text{mbar}$, (f) $2.3 \times 10^{-7}\,\text{mbar}$, (g) $3.7 \times 10^{-7}\,\text{mbar}$, (h) $6 \times 10^{-7}\,\text{mbar}$. The blue curve (f) corresponds to an effective cell temperature of $\sim 45\,°\text{C}$, which maximizes the 2D-MOT flux.

Appendix B

Saturation spectroscopy of the violet $4S_{1/2} \to 5P_{3/2}$ transition of K

Alkali atoms can be efficiently cooled and manipulated using laser light resonant with the D2 transition $nS_{1/2} \to nP_{3/2}$ of the atoms. These dipole transitions have large oscillator strengths and possess hyperfine sub-transitions which are closed transitions. Dipole transitions with a higher frequency (e.g. $nS_{1/2} \to (n+1)P_{3/2}$) generally have weaker oscillator strengths according to the Thomas-Reiche-Kuhn sum rule [279, 280]. Furthermore, because of the presence of intermediate transitions, the atom cannot be described as a two-level atom for these transitions. Nonetheless they can be used for several interesting applications: the higher transition frequency allows one to image atom clouds with a higher resolution (as that is limited by the wavelength of the imaging light). Furthermore, due to the smaller oscillator strengths the transitions have smaller line widths. Thus they can be used for efficient optical molasses cooling, as the minimum temperature of this cooling method is limited by the line width of the atomic transition used for cooling.

Potassium has a dipole transition in the violet frequency range, $4S_{1/2} \to 5P_{3/2}$, which has a wavelength of 404.414 nm and a narrow line width of $\Gamma^{(404)} = 2\pi \times 0.2$ MHz. Using this light for imaging could thus *a priori* yield a two-times higher resolution than light resonant with the D2 transition $4S_{1/2} \to 4P_{3/2}$, which has a wavelength of 766.490 nm and a line width of $\Gamma = 2\pi \times 6$ MHz (practical aspects for imaging with the violet transition have been discussed in Ref. [281]). The line width of the violet transition is ~ 30 times narrower than the line width of the infrared (IR) D2 transition, which leads to a very small Doppler temperature of only $T_{\mathrm{D}}^{(404)} = \hbar\Gamma^{(404)}/2k_{\mathrm{B}} = 4.8\,\mu$K as compared to $T_{\mathrm{D}}^{(766)} = 144\,\mu$K. These considerations motivated us to study the violet transition for potassium.

The recent development of laser diodes for the UV light regime makes laser sources available for such experiments. In particular, laser diodes emitting a wavelength of 404 nm are of low cost due to their mass production for the Blu-ray disc industry. We bought such a diode and set up an experiment for saturated absorption spectroscopy. The laser has been frequency stabilized and is ready to be used for future studies. The line width of the laser has been estimated from the noise of the error signal to be less than 500 kHz.

Absorption spectroscopy in alkali atom vapor cells using transitions to higher excited levels has been demonstrated in the past, in particular for rubidium [282, 283, 284], potassium [285, 286, 287] and lithium [288] for which the wavelength of the transitions $nS_{1/2} \rightarrow (n+1)P_{3/2}$ belong to a part of the spectrum, for which laser diodes exist (420 nm for Rb, 404 nm for K and 323 nm for Li). In the following we present our spectroscopy results and we compare them to our theoretical calculations. We calculate the populations of the different atomic levels of potassium in case of illumination with resonant violet light. We further calculate the cross section for the scattering of resonant violet light and compare it to the case of the IR transition. The calculations are presented for the case of potassium. We also give the results obtained for rubidium.

The laser diode used for our spectroscopy experiments (Sharp, ref. GH04020A2GE) emits 20 mW at a driving current of 60 mA and is operated at a temperature of $\sim$22 °C in external cavity configuration of Littrow-type (on the market, laser diodes with a higher output power of 120-140 mW are available (Mitsubishi, ref. ML320G2-11 or Nichia, ref. NDV4313)). The diode mount is homemade and identical to the ones we use for the other diode lasers of our setup. The external cavity has a length of $\sim$ 2 cm. The very small mode-hop free scanning range of $\sim$ 1 GHz is increased to $\sim$ 35 GHz by implementing the feed-forward technique [220]. The output beam of the diode is collimated with a broadband anti-reflection coated aspheric lens of 8 mm focal length (Thorlabs, ref. A240TM-A). The diffraction grating of the external cavity (Thorlabs, ref. GH13-36U) has a groove density of 3600 lines/mm with a diffraction efficiency of $\sim$ 10% into the first order for the used wavelength and polarisation.

The absorption spectroscopy is performed in a 5 cm long vapor cell (with natural isotopic abundances) which was heated to a temperature of $\sim$ 150 °C in order to achieve sufficient light absorption. The optical setup of the spectroscopy corresponds to the common pump-probe configuration: the laser beam is split into two parts of different intensities, which are sent through the cell from two opposite directions. The transmission profile of the weaker beam (the "probe" beam) is measured as the frequency of the laser is scanned. In order to compare the spectroscopy results for the violet transition with those for the IR transition, we implemented an identical setup for the IR laser, but with the absorption cell heated to a much lower temperature of $\sim$ 45 °C.

Figures B.1 and B.2 show typical transmission profiles for the IR and violet transitions, respectively, for different intensities of the pump beam. Each transmission profile consists of a Doppler profile with three sub-Doppler features. The two narrow peaks correspond to the transitions $S_{1/2}(F = 1) \to P_{3/2}$, $S_{1/2}(F = 2) \to P_{3/2}$ of ^{39}K and the dip to the corresponding crossover. The hyperfine structure of the excited states is not resolved. The sub-Doppler features for the violet transition have a significantly smaller width than those of the IR transition due to the approximately three-times smaller hyperfine structure of the excited state $5P_{3/2}$ [285]. The Doppler profiles of the violet transition have an approximately two-times larger width for than those of the IR transition, principally due to the larger transition frequency (Eq. (A.11)). In the following we are interested in the absorption characteristics of potassium with respect to resonant violet light. The figures demonstrate that the cross section for violet light scattering is smaller than for IR light scattering, since the relative light absorption for both frequencies is the same whereas the temperature of the vapor cell needed to be chosen much higher for the violet light in order to achieve this. Furthermore the figures show that much higher pump intensities (with respect to the corresponding saturation intensity) are required for the violet transition in order to obtain a reasonable contrast for the saturated absorption peaks.

In the following we characterize the violet transition and we explain the above observations. We calculate the populations of the different states of a potassium atom when it is illuminated with light resonant with the $4S_{1/2} \to 5P_{3/2}$ transition. In addition, we derive the cross section for scattering of the resonant violet light. Considering only the fine structure of the atom, seven energy levels need to be taken into account for the population analysis. An energy diagram of these levels is shown in Fig. B.3, with the corresponding spontaneous emission rates and emitted wavelengths given in Tab. B.1.

The excited state populations are given by the steady state solution of the system's master equation, that is, the equation of motion for the system's density matrix ρ. Within the rotating wave approximation and in the interaction picture, the master equation for the considered seven-level system reads [289, 290]

$$\dot{\rho} = -\frac{i}{\hbar}[H, \rho] + \gamma_{21}L_{21}\rho + \gamma_{31}L_{31}\rho + \gamma_{42}L_{42}\rho + \gamma_{43}L_{43}\rho + \gamma_{52}L_{52}\rho + \gamma_{53}L_{53}\rho$$
$$+ \gamma_{63}L_{63}\rho + \gamma_{71}L_{71}\rho + \gamma_{74}L_{74}\rho + \gamma_{75}L_{75}\rho + \gamma_{76}L_{76}\rho, \qquad (B.1)$$

184 Chapter B. Saturation spectroscopy of the violet $4S_{1/2} \to 5P_{3/2}$ transition of K

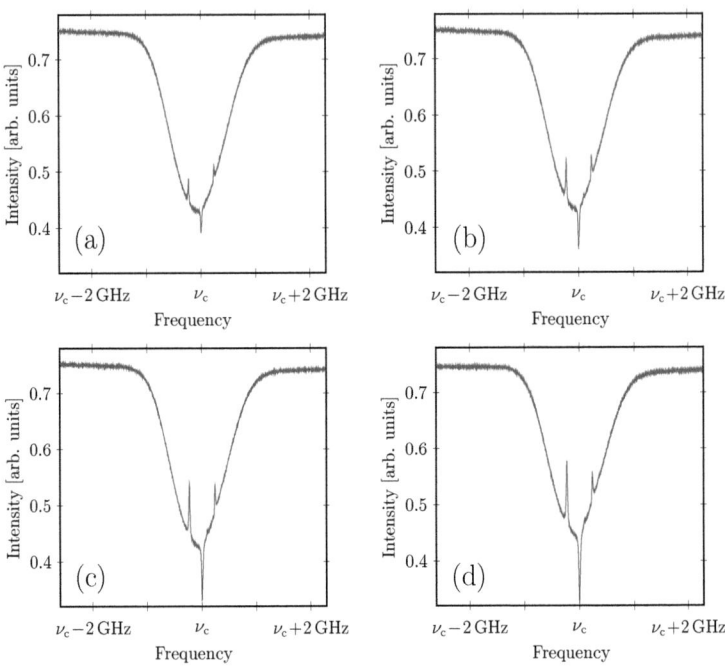

Figure B.1: Saturated absorption profile for the transition $4S_{1/2} \to 4P_{3/2}$ at 766.49 nm for different pump beam intensities I_{pump} and a fixed probe beam intensity of $I_{\text{probe}} \sim I_{\text{sat}}/6$, obtained with a 5 cm long vapor cell that was heated to $\sim 45°$C. (a) $I_{\text{pump}} \sim I_{\text{sat}}/3$, (b) $I_{\text{pump}} \sim I_{\text{sat}}$, (c) $I_{\text{pump}} \sim 4\,I_{\text{sat}}$, (d) $I_{\text{pump}} \sim 8\,I_{\text{sat}}$, with $I_{\text{sat}} \sim 1.75\,\text{mW/cm}^2$. In the graphs, ν_c denotes the frequency of the crossover transition $4S_{1/2}(F=1,2) \to 4P_{3/2}$ of ^{39}K.

with

$$L_{ij}\rho = \frac{1}{2}(2\sigma_{ji}\rho\sigma_{ij} - \sigma_{jj}\rho - \rho\sigma_{jj}), \tag{B.2}$$

and $\sigma_{ij} = |i\rangle\langle j|$ ($i,j \in \{1,2,3,4,5,6,7\}$), where H is the system's Hamiltonian given by

$$H = -\frac{\hbar}{2}\Omega\sigma_{17} - \frac{\hbar}{2}\Omega\sigma_{71}, \tag{B.3}$$

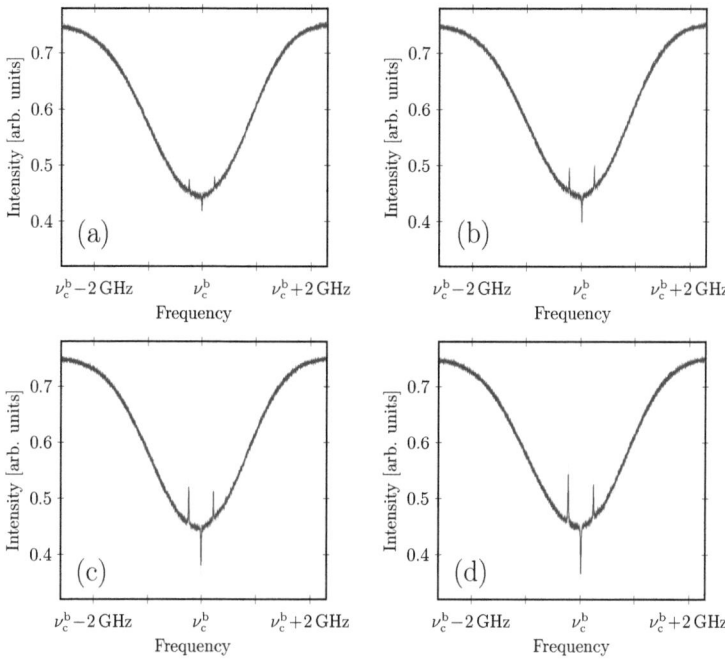

Figure B.2: Saturated absorption profile for the transition $4S_{1/2} \to 5P_{3/2}$ at 404.414 nm for different pump beam intensities I_{pump} and a fixed probe beam intensity of $I_{\text{probe}} \sim 3\, I_{\text{sat}}^{71}$, obtained with a 5 cm long vapor cell that was heated to $\sim 150°$C. (a) $I_{\text{pump}} \sim 16\, I_{\text{sat}}^{71}$, (b) $I_{\text{pump}} \sim 40\, I_{\text{sat}}^{71}$, (c) $I_{\text{pump}} \sim 80\, I_{\text{sat}}^{71}$, (d) $I_{\text{pump}} \sim 160\, I_{\text{sat}}^{71}$, with $I_{\text{sat}}^{71} \sim 0.39\,\text{mW/cm}^2$. In the graphs, ν_c^b denotes the frequency of the crossover transition $4S_{1/2}(F=1,2) \to 5P_{3/2}$ of ^{39}K.

with the on-resonant Rabi frequency Ω, that is defined as

$$\Omega = \frac{\mu_{71} E}{\hbar} = \gamma_{71}\sqrt{\frac{I}{2 I_{\text{sat}}^{71}}}, \tag{B.4}$$

where μ_{71} denotes the electric dipole moment of the atom in state $|7\rangle$, E the electric field of the incident light and I_{sat}^{71} the saturation intensity defined by

$$I_{\text{sat}}^{71} = \frac{2\pi^2 \hbar c \gamma_{71}}{3 \lambda_{71}^3} = 0.39\,\text{mW/cm}^2. \tag{B.5}$$

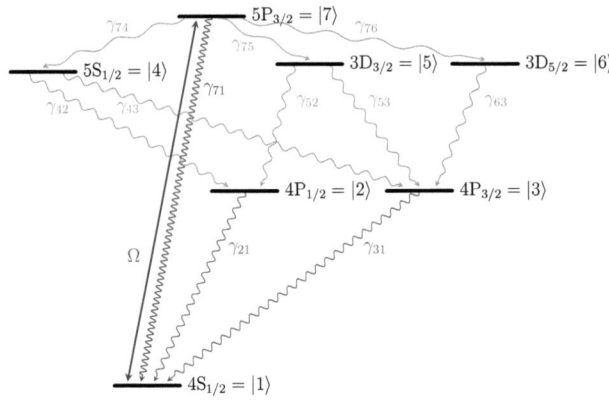

Figure B.3: Fine-structure energy levels of potassium relevant for the violet transition $4S_{1/2} \to 5P_{3/2}$ (energy level splittings to scale).

$\gamma_{21} = 3.82 \times 10^7 \, \text{s}^{-1}$	$\lambda_{21} = 769.896 \, \text{nm}$
$\gamma_{31} = 3.87 \times 10^7 \, \text{s}^{-1}$	$\lambda_{31} = 766.490 \, \text{nm}$
$\gamma_{42} = 7.90 \times 10^6 \, \text{s}^{-1}$	$\lambda_{42} = 1243.224 \, \text{nm}$
$\gamma_{43} = 1.56 \times 10^7 \, \text{s}^{-1}$	$\lambda_{43} = 1252.211 \, \text{nm}$
$\gamma_{52} = 2.20 \times 10^7 \, \text{s}^{-1}$	$\lambda_{52} = 1169.021 \, \text{nm}$
$\gamma_{53} = 4.34 \times 10^6 \, \text{s}^{-1}$	$\lambda_{53} = 1176.962 \, \text{nm}$
$\gamma_{63} = 2.59 \times 10^7 \, \text{s}^{-1}$	$\lambda_{63} = 1177.283 \, \text{nm}$
$\gamma_{71} = 1.24 \times 10^6 \, \text{s}^{-1}$	$\lambda_{71} = 404.414 \, \text{nm}$
$\gamma_{74} = 4.60 \times 10^6 \, \text{s}^{-1}$	$\lambda_{74} = 2707.395 \, \text{nm}$
$\gamma_{75} = 1.50 \times 10^5 \, \text{s}^{-1}$	$\lambda_{75} = 3141.541 \, \text{nm}$
$\gamma_{76} = 1.40 \times 10^6 \, \text{s}^{-1}$	$\lambda_{76} = 3139.265 \, \text{nm}$

Table B.1: Scattering rates and emitted wavelengths obtained from Ref. [226]

Substituting Eq. (B.3) into Eq. (B.1) yields the optical Bloch equations

$$\dot{\rho}_{11} = \gamma_{21}\rho_{22} + \gamma_{31}\rho_{33} + \gamma_{71}\rho_{77} + \frac{i\Omega}{2}(\rho_{71} - \rho_{17}), \tag{B.6}$$

$$\dot{\rho}_{22} = -\gamma_{21}\rho_{22} + \gamma_{42}\rho_{44} + \gamma_{52}\rho_{55}, \tag{B.7}$$

$$\dot{\rho}_{33} = -\gamma_{31}\rho_{33} + \gamma_{43}\rho_{44} + \gamma_{53}\rho_{55} + \gamma_{63}\rho_{66}, \tag{B.8}$$

$$\dot{\rho}_{44} = -\gamma_{42}\rho_{44} - \gamma_{43}\rho_{44} + \gamma_{74}\rho_{77}, \tag{B.9}$$

$$\dot{\rho}_{55} = -\gamma_{52}\rho_{55} - \gamma_{53}\rho_{55} + \gamma_{75}\rho_{77}, \tag{B.10}$$

$$\dot{\rho}_{66} = -\gamma_{63}\rho_{66} + \gamma_{76}\rho_{77}, \tag{B.11}$$

$$\dot{\rho}_{77} = -\gamma_{71}\rho_{77} - \gamma_{74}\rho_{77} - \gamma_{75}\rho_{77} - \gamma_{76}\rho_{77} + \frac{i\Omega}{2}(\rho_{17} - \rho_{71}), \tag{B.12}$$

$$\dot{\rho}_{17} = -\frac{\gamma_{71}}{2}\rho_{17} - \frac{\gamma_{74}}{2}\rho_{17} - \frac{\gamma_{75}}{2}\rho_{17} - \frac{\gamma_{76}}{2}\rho_{17} + \frac{i\Omega}{2}(\rho_{77} - \rho_{11}), \tag{B.13}$$

$$\dot{\rho}_{71} = -\frac{\gamma_{71}}{2}\rho_{71} - \frac{\gamma_{74}}{2}\rho_{71} - \frac{\gamma_{75}}{2}\rho_{71} - \frac{\gamma_{76}}{2}\rho_{71} + \frac{i\Omega}{2}(\rho_{11} - \rho_{77}), \tag{B.14}$$

where $\rho_{ij} = \langle i|\rho|j\rangle$. In the stationary case, $\dot{\rho}_{ij} = 0$. Utilizing that the trace of the density matrix is unity,

$$\rho_{11} + \rho_{22} + \rho_{33} + \rho_{44} + \rho_{55} + \rho_{66} + \rho_{77} = 1, \tag{B.15}$$

allows solving the system of Eqs. (B.6-B.15) for the stationary case. One obtains the following steady state populations of the different states:

$$\rho_{11} = 0.414 \times \frac{s + 71.03}{s + 29.41}, \tag{B.16}$$

$$\rho_{22} = 0.018 \times \frac{s}{s + 29.41}, \tag{B.17}$$

$$\rho_{33} = 0.048 \times \frac{s}{s + 29.41}, \tag{B.18}$$

$$\rho_{44} = 0.081 \times \frac{s}{s + 29.41}, \tag{B.19}$$

$$\rho_{55} = 0.002 \times \frac{s}{s + 29.41}, \tag{B.20}$$

$$\rho_{66} = 0.022 \times \frac{s}{s + 29.41}, \tag{B.21}$$

$$\rho_{77} = 0.414 \times \frac{s}{s + 29.41}, \tag{B.22}$$

with $s = I/I_{\text{sat}}^{71}$. Figure B.4 shows the level populations for the limit of very large intensity $(I \gg I_{\text{sat}}^{71})$ and for the particular intensity $I = 22\, I_{\text{sat}}^{71}$.

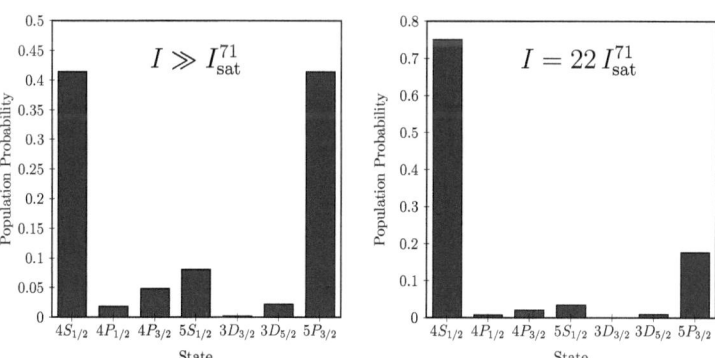

Figure B.4: State population probability for a K-atom that is driven on the violet transition $4S_{1/2} \to 5P_{3/2}$ with (left) very large intensity $(I \gg I_{\text{sat}}^{71})$ and (right) the particular intensity $I = 22\, I_{\text{sat}}^{71}$, for which the atom populates the ground state $4S_{1/2}$ with a probability of 75%. In a simple two-level atom approximation, this population probability would be expected for $I = I_{\text{sat}}^{71}$.

For the IR transition the hyperfine peaks of the recorded transmission profiles have a reasonable size for a pump intensity $I = I_{\text{sat}}$ (see Fig. B.1 (b)). For this intensity, 75% of the atoms populate the ground state. For the violet transition, hyperfine peaks of a comparable size are thus expected for a pump intensity which yields the same ground state population, *i.e.* $\rho_{11} = 0.75$. According to Eq. (B.16) this is equivalent to

$$I = 22\, I_{\text{sat}}^{71}. \tag{B.23}$$

Thus, a much larger pump intensity is required for the violet transition in order to induce a relative ground state population of 75%. The difference with respect to the IR transition is a consequence of the additional decay channels present in the case of excitation with violet light. According to Eq. (B.23), we expect the contrast of the hyperfine peaks to be the same for the IR transition with $I = I_{\text{sat}}$ and the violet transition with $I = 22\, I_{\text{sat}}^{71}$. Figures B.1 (b) and B.2 (b) show that the contrast for the IR transition with $I = I_{\text{sat}}$ is obtained for the violet transition with $I = 40\, I_{\text{sat}}^{71}$. This value is thus by a factor of about two larger than the calculated value. The difference might be due to the high intensity of the violet probe beam which was used in the experiment, which is causing significant stimulated emission and thus leads to a decrease of the peak contrast. Another reason might be that the hyperfine and Zeeman structures as well as the polarization of the light had been neglected in the theoretical calculation. However, the intensity profiles of the laser beams being highly inhomogeneous and the overlap of the pump and probe beams being difficult to control, a precise quantitative comparison is difficult.

In order to obtain a reasonable absorption of the probe beam by the vapor we observed in our spectroscopy experiment that a much higher vapor pressure is required for the violet transition (as we need to heat the vapor cell to $\sim 150\,^\circ\text{C}$, compared to $\sim 45\,^\circ\text{C}$ for the IR transition). This indicates that the absorption cross section for scattering with violet resonant light is smaller than that for IR resonant light. In the following we calculate both cross sections for a comparison. The absorption cross section σ is defined as the ratio of the total scattered power P_{sc} and the incoming intensity I:

$$\sigma = \frac{P_{\text{sc}}}{I}, \tag{B.24}$$

where P_{sc} for the violet transition is given by (using the notation $\nu_{ij} = c/\lambda_{ij}$)

$$\begin{aligned}P_{\text{sc}} &= h\left[\rho_{77}(\gamma_{71}\nu_{71} + \gamma_{74}\nu_{74} + \gamma_{75}\nu_{75} + \gamma_{76}\nu_{76}) + \rho_{66}\gamma_{63}\nu_{63} + \rho_{55}(\gamma_{52}\nu_{52} + \gamma_{53}\nu_{53})\right.\\ &\quad \left. + \rho_{44}(\gamma_{42}\nu_{42} + \gamma_{43}\nu_{43}) + \rho_{33}\gamma_{31}\nu_{31} + \rho_{22}\gamma_{21}\nu_{21}\right]\\ &= \frac{s}{s + 29.41} \times 1.503 \times 10^{-12}\,\text{W}.\end{aligned} \quad (B.25)$$

Substituting Eq. (B.25) into Eq. (B.24) yields

$$\sigma^{(404)} = \frac{1.503 \times 10^{-12}\,\text{W}}{I + 29.41 I_{\text{sat}}^{71}}. \quad (B.26)$$

For $I \ll I_{\text{sat}}^{71}$, $\sigma^{(404)}$ takes the value

$$\sigma^{(404)} = 1.31 \times 10^{-14}\,\text{m}^2. \quad (B.27)$$

For the IR transition at 766.490 nm the scattering cross section is given by that of a two-level atom,

$$\sigma^{(766)} = \frac{3 \times (766.490 \times 10^{-9})^2}{2\pi} = 2.805 \times 10^{-13}\,\text{m}^2, \quad (B.28)$$

such that

$$\sigma^{(404)} = \frac{1}{21.4}\sigma^{(766)}. \quad (B.29)$$

Thus the scattering cross section of violet resonant light is 21.4 times smaller than that of IR resonant light. The suppression factor has two contributions, one coming from the higher transition frequency of the violet transition (a two-level approximation would yield $\sigma_{2L}^{(404)} = \sigma^{(766)}/3.6$) and the other from the presence of the additional decay channels.

We also calculated the scattering cross section for the case of a rubidium atom illuminated by violet light resonant with the transition $5S_{1/2} \to 6P_{3/2}$. This transition has a wavelength of 420.3 nm and a line width of $\Gamma^{(420)} = 2\pi \times 1.8\,\text{MHz}$ [226, 291]. The scattering cross section is found to be $\sigma^{(420)} = \sigma^{(780)}/10.4$, where $\sigma^{(780)} = 2.9 \times 10^{-13}\,\text{m}^2$ is the cross section for scattering of light resonant with the D2 transition $5S_{1/2} \to 5P_{3/2}$.

In conclusion, we have built a diode laser which emits 20 mW of narrow-band light of 404 nm wavelength, which is resonant with the violet transition $4S_{1/2} \to 5P_{3/2}$ of potassium. We have used this laser to perform saturated absorption spectroscopy in a vapor cell. High vapor cell temperatures were required to obtain sufficient absorption of the violet light resulting from the small associated scattering cross section. We found that the violet atomic transition saturates for an intensity of about an order of magnitude higher than would be expected from the two-level atom approximation, due to the additional intermediate atomic energy levels. We verified our experimental findings by theoretical calculations of the population of the different atomic energy levels under resonant violet light illumination and the corresponding scattering cross section.

The built laser system can be used in future experiments for high-resolution imaging or for efficient optical molasses cooling of potassium atoms. The higher frequency of the violet transition as compared to the common IR transition would allow a two-times better imaging resolution. Optical molasses cooling would be efficient, since the narrow line width of the violet transition leads to a thirty-times smaller Doppler temperature of only $\sim 4.8\,\mu$K as compared to $\sim 144\,\mu$K for the IR transition. Even sub-Doppler cooling might be possible using the violet transition. The several cascade-like decay channels might, however, prevent the necessary polarization gradients from building up and thus limit the efficiency of sub-Doppler cooling.

Appendix C

Engineering drawings

C.1 Octagonal cell

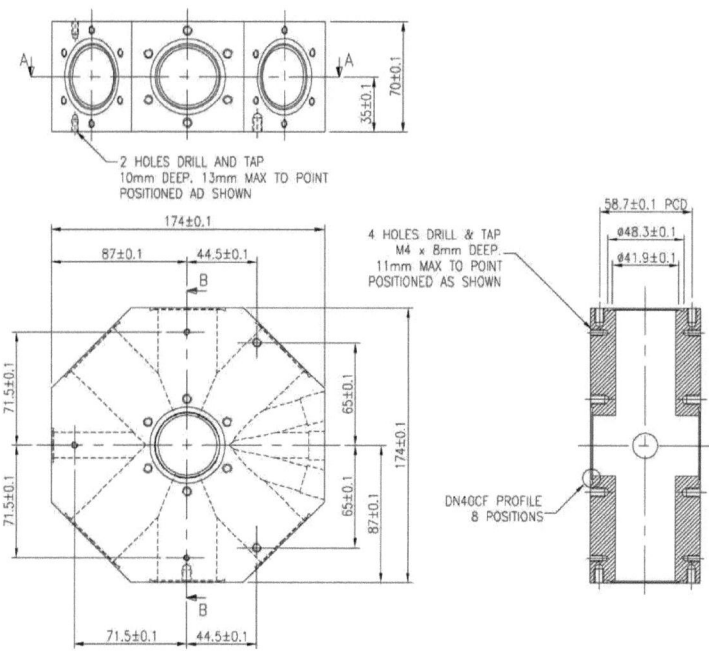

Figure C.1: Engineering drawings of the vacuum octagonal cell (manufactured by Caburn-MDC Europe). Indicated dimensions are in units of mm.

C.2 Science cell

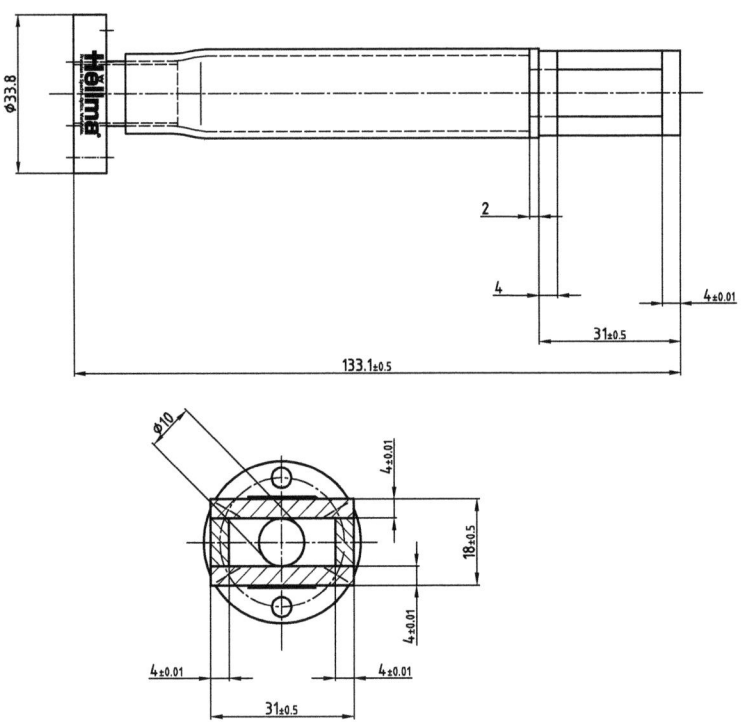

Figure C.2: Engineering drawings of the vacuum science cell (manufactured by Hellma GmbH). Indicated dimensions are in units of mm. Additional specifications are given in Sec. 2.2.1.

C.3 Tapered amplifier support for potassium

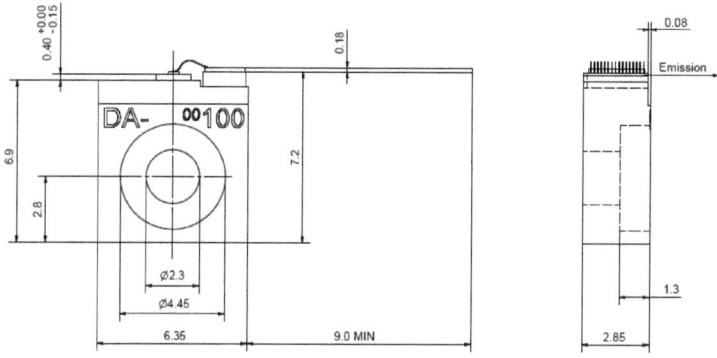

Figure C.3: Engineering drawings of the tapered amplifier chip for potassium (Eagleyard, ref. EYP-RWE-0790-0400-0750-SOT03-0000), to be mounted on the homemade support whose design is detailed in the subsequent figures.

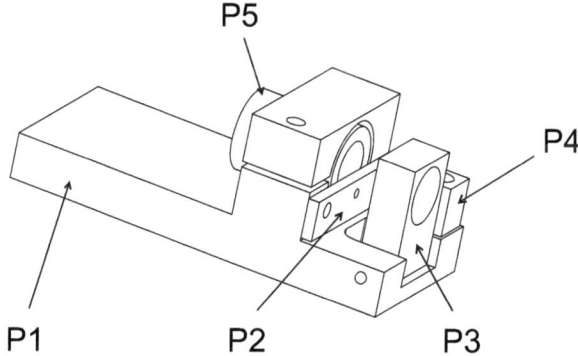

Figure C.4: Schematic of the support for the tapered amplifier chip. The different parts include: a base (P1), a laser chip support (P2) and mounts for the output collimation lens (P3), the isolated blade connectors (P4) and the input collimation lens (P5), which are detailed in the subsequent figures. The chosen materials are: duralumin (P1), copper (P2, P3, P5) and teflon (P4).

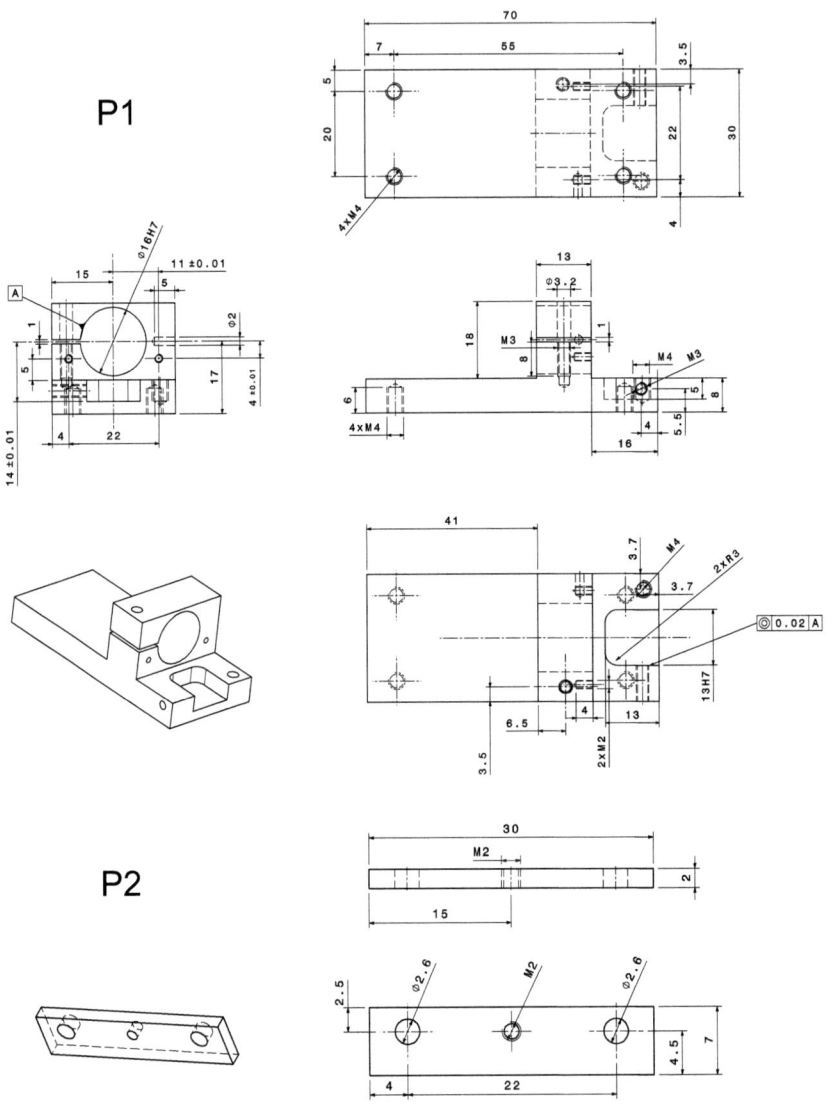

Figure C.5: Engineering drawings of the base (P1) and the laser chip mount (P2). Required precision: ±0.1 mm, except specified otherwise.

C.3. Tapered amplifier support for potassium

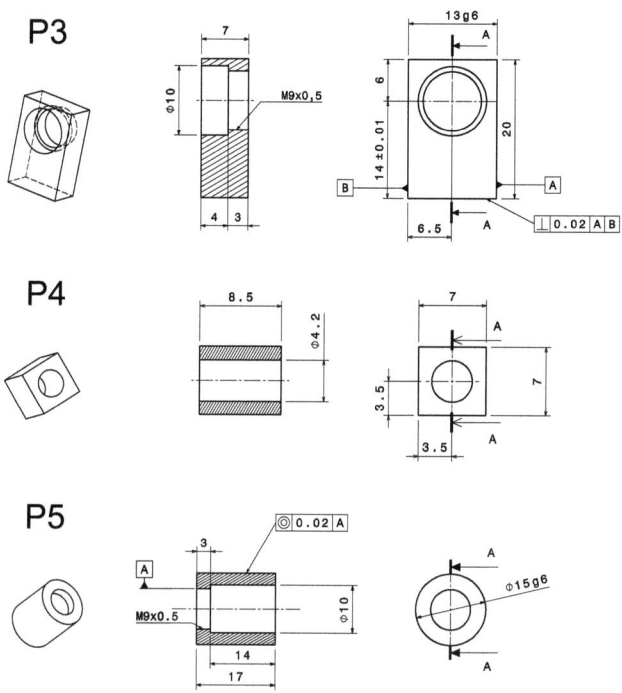

Figure C.6: Engineering drawings of the mounts for the output collimation lens (P3), the isolated blade connectors (P4) and the input collimation lens (P5). Required precision: ±0.1 mm, except specified otherwise.

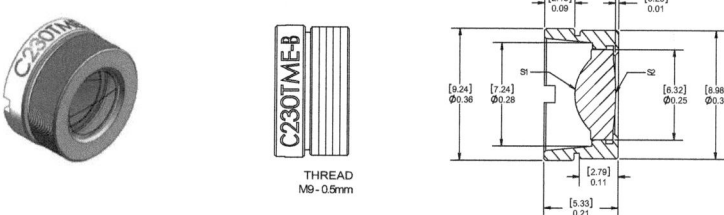

Figure C.7: Engineering drawings of the collimation lens socket (Thorlabs, ref. C230TME-B), to be screwed into the mounts P3 and P5 (entirely).

C.4 Tapered amplifier support for lithium

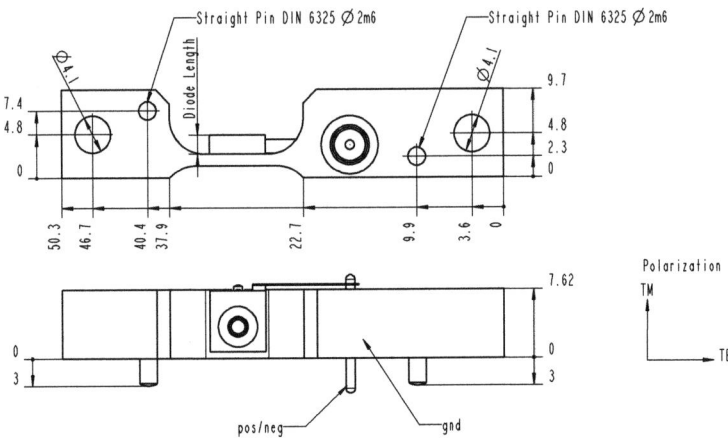

Figure C.8: Engineering drawings of the tapered amplifier chip for lithium (Toptica, ref. TA-670-0500-5), to be mounted on the homemade support whose design is detailed in the subsequent figures.

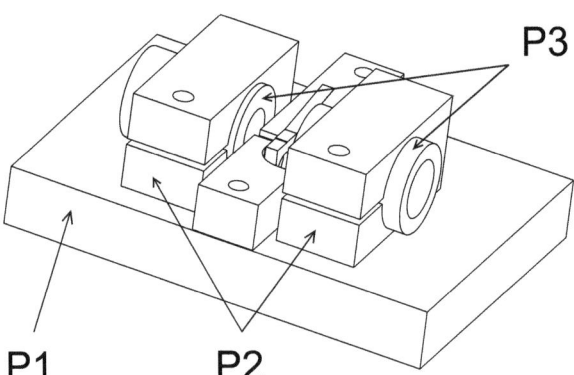

Figure C.9: Schematic of the support for the tapered amplifier chip. The different parts include: a base (P1), collimation tube supports (P2) and the collimation tubes (P3), which are detailed in the subsequent figures. The chosen material for all pieces is duralumin.

C.4. Tapered amplifier support for lithium

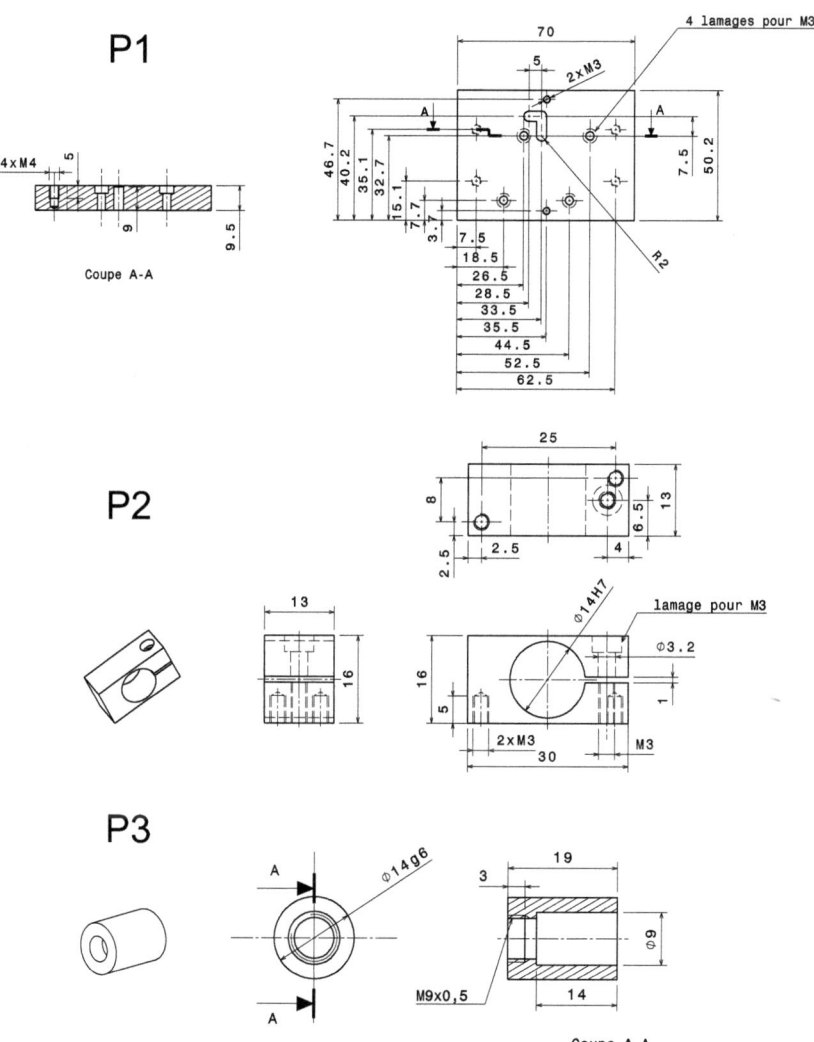

Figure C.10: Engineering drawings of the base (P1), the collimation tube supports (P2) and the collimation tubes (P3). Required precision: ±0.1 mm, except specified otherwise. The dimensions of the collimation lenses are given in Fig. C.7.

C.5 2D-MOT vacuum parts

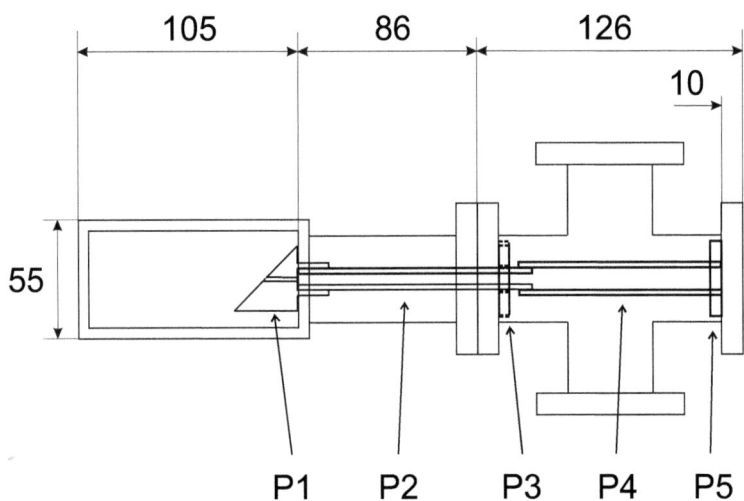

Figure C.11: Schematic of the 2D-MOT vacuum parts showing the glass cell, a standard CF40 cross and homemade stainless steel pieces, including a polished mirror (P1), two differential pumping tubes (P2, P4) and two support rings (P3, P5), which are detailed in the subsequent figures. The inner diameter of the glass-to-metal junction of the glass cell is 32 mm and of the CF40 cross 34.9 mm, the thickness of the walls of the glass cell is 5 mm. The total differential pumping tube has a diameter of 2 mm over 20 mm, 5 mm over 106 mm and 10 mm over 86 mm.

C.5. 2D-MOT vacuum parts

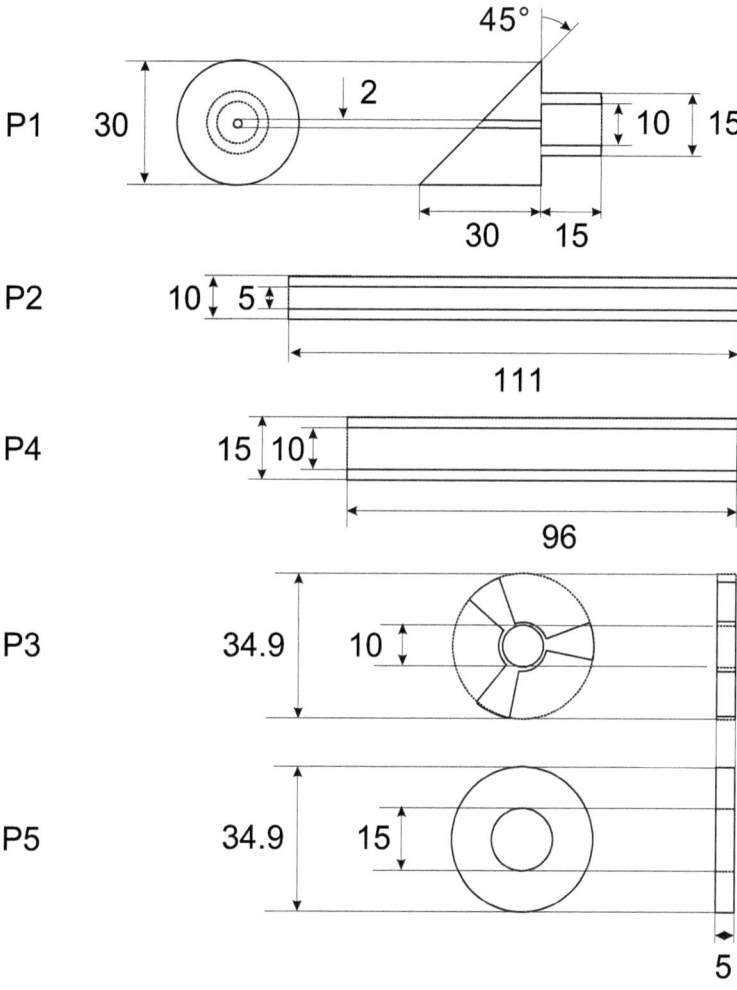

Figure C.12: Engineering drawings of the homemade pieces, *i.e.* the mirror (P1), the first differential pumping tube (P2), the first support ring (P3), the second differential pumping tube (P4) and the second support ring (P5). The material is of type 304.

Appendix D

Publications

The work presented in this book has been published in the following peer-reviewed journal articles:

- A. Ridinger and N. Davidson, *Particle motion in rapidly oscillating potentials: The role of the potential's initial phase*, Phys. Rev. A **76**, 013421 (2007)

- A. Ridinger and C. Weiss, *Manipulation of quantum particles in rapidly oscillating potentials by inducing phase hops*, Phys. Rev. A **79**, 013414 (2009)

- A. Ridinger, S. Chaudhuri, T. Salez, U. Eismann, D. R. Fernandes, K. Magalhães, D. Wilkowski, C. Salomon and F. Chevy, *Large atom number dual-species magneto-optical trap for fermionic ^{6}Li and ^{40}K atoms*, Eur. Phys. J. D (2011)

- A. Ridinger, S. Chaudhuri, T. Salez, D. R. Fernandes, N. Bouloufa, O. Dulieu, C. Salomon and F. Chevy, *Photoassociative creation of ultracold heteronuclear $^6Li^{40}K^*$ molecules*, Eur. Phys. Lett. (2011), *in press*, preprint: *arXiv:1108.0618*

Bibliography

[1] A. Einstein. Quantentheorie des idealen einatomigen Gases. *Sitzungsber. Preuss. Akad. Wiss.*, 3:18, 1925.

[2] S. Bose. Plancks Gesetz und Lichtquantenhypothese. *Zeitschrift für Physik*, 26:178, 1924.

[3] P. Kapitza. Viscosity of liquid helium below the λ-point. *Nature*, 141:3558, 1938.

[4] P. Kapitza. Flow of liquid helium II. *Nature*, 141:75, 1938.

[5] F. London. On the Bose-Einstein condensation. *Phys. Rev.*, 54:947, 1938.

[6] M. H. Anderson, J. R. Ensher, M. R. Matthews, C. E. Wieman, and E. A. Cornell. Observation of Bose-Einstein condensation in a dilute atomic vapor. *Science*, 269:198, 1995.

[7] C. C. Bradley, C. A. Sackett, J. J. Tollett, and R. G. Hulet. Evidence of Bose-Einstein condensation in an atomic gas with attractive interactions. *Phys. Rev. Lett.*, 75:1687, 1995.

[8] K. B. Davis, M.-O. Mewes, M. R. Andrews, N. J. van Druten, D. S. Durfee, D. M. Kurn, and W. Ketterle. Bose-Einstein condensation in a gas of sodium atoms. *Phys. Rev. Lett.*, 75:3969, 1995.

[9] W. Ketterle, D. S. Durfee, and D. M. Stamper-Kurn. Making, probing and understanding Bose-Einstein condensates. In M. Inguscio, S. Stringari, and C. E. Wieman, editors, *Proceedings of the International School of Physics Enrico Fermi, course CXL*, page 67, 1999.

[10] A. J. Leggett. Bose-Einstein condensation in the alkali gases: Some fundamental concepts. *Rev. Mod. Phys.*, 73:307, 2001.

[11] L. Pitaevski and S. Stringari, editors. *Bose-Einstein Condensation*. Oxford University Press, Oxford, UK, 2003.

[12] C. J. Pethick and H. Smith, editors. *Bose-Einstein Condensation in Dilute Gases*, 2nd ed. Cambridge University Press, Cambridge, UK, 2008.

[13] M. R. Andrews, C. G. Townsend, H.-J. Miesner, D. S. Durfee, D. M. Kurn, and W. Ketterle. Observation of interference between two Bose-Einstein condensates. *Science*, 275:637, 1997.

[14] K. W. Madison, F. Chevy, W. Wohlleben, and J. Dalibard. Vortex formation in a stirred Bose-Einstein condensate. *Phys. Rev. Lett.*, 84:806, 2000.

[15] J. R. Abo-Shaeer, C. Raman, J. M. Vogels, and W. Ketterle. Observation of vortex lattices in Bose-Einstein condensates. *Science*, 292:476, 2001.

[16] J. Denschlag, J. E. Simsarian, D. L. Feder, C. W. Clark, L. A. Collins, J. Cubizolles, L. Deng, E.W. Hagley, K. Helmerson, W. P. Reinhardt, S. L. Rolston, B. I. Schneider, and W. D. Phillips. Generating solitons by phase engineering of a Bose-Einstein condensate. *Science*, 287:97, 2000.

[17] S. Burger, K. Bongs, S. Dettmer, W. Ertmer, K. Sengstock, A. Sanpera, G. V. Shlyapnikov, and M. Lewenstein. Dark solitons in Bose-Einstein condensates. *Phys. Rev. Lett.*, 83:5198, 1999.

[18] L. Khaykovich, F. Schreck, G. Ferrari, T. Bourdel, J. Cubizolles, L. D. Carr, Y. Castin, and C. Salomon. Formation of a matter-wave bright soliton. *Science*, 296:1290, 2002.

[19] K. E. Strecker, G. B. Partridge, A. G. Truscott, and R. G. Hulet. Formation and propagation of matter-wave soliton trains. *Nature*, 417:150, 2002.

[20] M. Greiner, O. Mandel, T. Esslinger, T.W. Hänsch, and I. Bloch. Quantum phase transition from a superfluid to a Mott insulator in a gas of ultracold atoms. *Nature*, 415:39, 2002.

[21] B. DeMarco and D. S. Jin. Onset of Fermi degeneracy in a trapped atomic gas. *Science*, 285:1703, 1999.

[22] A. G. Truscott, K. E. Strecker, W. I. McAlexander, G. B. Partridge, and R. G. Hulet. Observation of Fermi pressure in a gas of trapped atoms. *Science*, 291:2570, 2001.

[23] F. Schreck, L. Khaykovich, K. L. Corwin, G. Ferrari, T. Bourdel, J. Cubizolles, and C. Salomon. Quasipure Bose-Einstein condensate immersed in a Fermi sea. *Phys. Rev. Lett.*, 87:080403, 2001.

[24] J. M. McNamara, T. Jeltes, A. S. Tychkov, W. Hogervorst, and W. Vassen. Degenerate Bose-Fermi mixture of metastable atoms. *Phys. Rev. Lett.*, 97:080404, 2006.

[25] T. Fukuhara, Y. Takasu, M. Kumakura, and Y. Takahashi. Degenerate Fermi gases of Ytterbium. *Phys. Rev. Lett.*, 98:030401, 2007.

[26] B. J. DeSalvo, M. Yan, P. G. Mickelson, Y. N. Martinez de Escobar, and T. C. Killian. Degenerate Fermi gas of ^{87}Sr. *Phys. Rev. Lett.*, 105:030402, 2010.

[27] M. K. Tey, S. Stellmer, R. Grimm, and F. Schreck. Double-degenerate Bose-Fermi mixture of strontium. *Phys. Rev. A*, 82:011608, 2010.

[28] Z. Hadzibabic, S. Gupta, C. A. Stan, C. H. Schunck, M. W. Zwierlein, K. Dieckmann, and W. Ketterle. Fiftyfold improvement in the number of quantum degenerate fermionic atoms. *Phys. Rev. Lett.*, 91:160401, 2003.

[29] S. Nascimbène, N. Navon, K. J. Jiang, F. Chevy, and C. Salomon. Exploring the thermodynamics of a universal Fermi gas. *Nature*, 463:1057, 2010.

[30] M. Inguscio, W. Ketterle, and C. Salomon, editors. *Proceedings of the International School of Physics Enrico Fermi on Ultracold Fermi Gases*, Bologna, Italy, 2006. Societa Italiana di Fisica.

[31] E. Timmermans. Degenerate fermion gas heating by hole creation. *Phys. Rev. Lett.*, 87:240403, 2001.

[32] L. D. Carr, T. Bourdel, and Y. Castin. Limits of sympathetic cooling of fermions by zero-temperature bosons due to particle losses. *Phys. Rev. A*, 69:033603, 2004.

[33] S. R. Granade, M. E. Gehm, K. M. O'Hara, and J. E. Thomas. All-optical production of a degenerate Fermi gas. *Phys. Rev. Lett.*, 88:120405, 2002.

[34] S. Jochim, M. Bartenstein, A. Altmeyer, G. Hendl, C. Chin, J. Hecker Denschlag, and R. Grimm. Pure gas of optically trapped molecules created from fermionic atoms. *Phys. Rev. Lett.*, 91:240402, 2003.

[35] C. Silber, S. Günther, C. Marzok, B. Deh, Ph. W. Courteille, and C. Zimmermann. Quantum-degenerate mixture of fermionic Lithium and bosonic Rubidium gases. *Phys. Rev. Lett.*, 95:170408, 2005.

[36] G. Roati, F. Riboli, G. Modugno, and M. Inguscio. Fermi-Bose quantum degenerate ^{40}K-^{87}Rb mixture with attractive interaction. *Phys. Rev. Lett.*, 89:150403, 2002.

[37] S. Inouye, J. Goldwin, M. L. Olsen, C. Ticknor, J. L. Bohn, and D. S. Jin. Observation of heteronuclear Feshbach resonances in a mixture of bosons and fermions. *Phys. Rev. Lett.*, 93:183201, 2004.

[38] Michael Köhl, Henning Moritz, Thilo Stöferle, Kenneth Günter, and Tilman Esslinger. Fermionic atoms in a three dimensional optical lattice: Observing Fermi surfaces, dynamics, and interactions. *Phys. Rev. Lett.*, 94(8):080403, Mar 2005.

[39] C. Ospelkaus, S. Ospelkaus, L. Humbert, P. Ernst, K. Sengstock, and K. Bongs. Ultracold heteronuclear molecules in a 3d optical lattice. *Phys. Rev. Lett.*, 97:120402, 2006.

[40] T. Rom, T. Best, D. van Oosten, U. Schneider, S. Fölling, B. Paredes, and I. Bloch. Free fermion antibunching in a degenerate atomic Fermi gas released from an optical lattice. *Nature*, 444:733, 2006.

[41] M. Taglieber, A.-C. Voigt, T. Aoki, T. W. Hänsch, and K. Dieckmann. Quantum degenerate two-species Fermi-Fermi mixture coexisting with a Bose-Einstein condensate. *Phys. Rev. Lett.*, 100:010401, 2008.

[42] C.-H. Wu, I. Santiago, J. W. Park, P. Ahmadi, and M. W. Zwierlein. Strongly interacting isotopic Bose-Fermi mixture immersed in a Fermi sea. *Phys. Rev. A*, 84:011601, 2011.

[43] F. M. Spiegelhalder, A. Trenkwalder, D. Naik, G. Kerner, E. Wille, G. Hendl, F. Schreck, and R. Grimm. All-optical production of a degenerate mixture of ^{6}Li and ^{40}K and creation of heteronuclear molecules. *Phys. Rev. A*, 81:043637, 2010.

[44] T. Tiecke. *Ph.D. thesis*. University of Amsterdam, 2009.

[45] B. DeMarco, S. B. Papp, and D. S. Jin. Pauli blocking of collisions in a quantum degenerate atomic Fermi gas. *Phys. Rev. Lett.*, 86:5409, 2001.

[46] H. Feshbach. Unified theory of nuclear reactions. *Ann. Phys. (N.Y.)*, 5:357, 1958.

[47] U. Fano. Effects of configuration interaction on intensities and phase shifts. *Phys. Rev.*, 124:1866, 1961.

[48] E. Tiesinga, B. J. Verhaar, and H. T. C. Stoof. Threshold and resonance phenomena in ultracold ground-state collisions. *Phys. Rev. A*, 47:4114, 1993.

[49] S. Inouye, M. R. Andrews, J. Stenger, H.-J. Miesner, D. M. Stamper-Kurn, and W. Ketterle. Observation of Feshbach resonances in a Bose-Einstein condensate. *Nature*, 392:151, 1998.

[50] K. M. O'Hara, S. L. Hemmer, S. R. Granade, M. E. Gehm, J. E. Thomas, V. Venturi, E. Tiesinga, and C. J. Williams. Measurement of the zero crossing in a Feshbach resonance of fermionic ^{6}Li. *Phys. Rev. A*, 66:041401, 2002.

[51] K. Dieckmann, C. A. Stan, S. Gupta, Z. Hadzibabic, C. H. Schunck, and W. Ketterle. Decay of an ultracold fermionic lithium gas near a Feshbach resonance. *Phys. Rev. Lett.*, 89:203201, 2002.

[52] S. Jochim, M. Bartenstein, G. Hendl, J. Hecker Denschlag, R. Grimm, A. Mosk, and M. Weidemüller. Magnetic field control of elastic scattering in a cold gas of fermionic lithium atoms. *Phys. Rev. Lett.*, 89:273202, 2002.

[53] T. Loftus, C. A. Regal, C. Ticknor, J. L. Bohn, and D. S. Jin. Resonant control of elastic collisions in an optically trapped Fermi gas of atoms. *Phys. Rev. Lett.*, 88:173201, 2002.

[54] C. A. Regal and D. S. Jin. Measurement of positive and negative scattering lengths in a Fermi gas of atoms. *Phys. Rev. Lett.*, 90:230404, 2003.

[55] C. A. Regal, C. Ticknor, J. L. Bohn, and D. S. Jin. Tuning p-wave interactions in an ultracold Fermi gas of atoms. *Phys. Rev. Lett.*, 90:053201, 2003.

[56] L. Cooper. Bound electron pairs in a degenerate Fermi gas. *Phys. Rev.*, 104:1189, 1957.

[57] J. Bardeen, L. Cooper, and J. Schrieffer. Theory of superconductivity. *Phys. Rev.*, 108:1175, 1957.

[58] C. A. Regal, C. Ticknor, J. L. Bohn, and D. S. Jin. Creation of ultracold molecules from a Fermi gas of atoms. *Nature*, 424:47, 2003.

[59] J. Cubizolles, T. Bourdel, S. J. J. M. F. Kokkelmans, G. V. Shlyapnikov, and C. Salomon. Production of long-lived ultracold Li_2 molecules from a Fermi gas. *Phys. Rev. Lett.*, 91:240401, 2003.

[60] K. E. Strecker, G. B. Partridge, and R. G. Hulet. Conversion of an atomic Fermi gas to a long-lived molecular Bose gas. *Phys. Rev. Lett.*, 91:080406, 2003.

[61] C. A. Regal, M. Greiner, and D. S. Jin. Lifetime of molecule-atom mixtures near a Feshbach resonance in ^{40}K. *Phys. Rev. Lett.*, 92:083201, 2004.

[62] D. S. Petrov, C. Salomon, and G. V. Shlyapnikov. Weakly bound dimers of fermionic atoms. *Phys. Rev. Lett.*, 93:090404, 2004.

[63] M. Greiner, C. A. Regal, and D. S. Jin. Emergence of a molecular Bose-Einstein condansate from a Fermi gas. *Nature*, 426:537, 2003.

[64] M. W. Zwierlein, C. A. Stan, C. H. Schunck, S. M. F. Raupach, S. Gupta, Z. Hadzibabic, and W. Ketterle. Observation of Bose-Einstein condensation of molecules. *Phys. Rev. Lett.*, 91:250401, 2003.

[65] C. A. Regal, M. Greiner, and D. S. Jin. Observation of resonance condensation of fermionic atom pairs. *Phys. Rev. Lett.*, 92:040403, 2004.

[66] M. W. Zwierlein, C. A. Stan, C. H. Schunck, S. M. F. Raupach, A. J. Kerman, and W. Ketterle. Condensation of pairs of fermionic atoms near a Feshbach resonance. *Phys. Rev. Lett.*, 92:120403, 2004.

[67] M. Bartenstein, A. Altmeyer, S. Riedl, S. Jochim, C. Chin, J. Hecker Denschlag, and R. Grimm. Crossover from a molecular Bose-Einstein condensate to a degenerate Fermi gas. *Phys. Rev. Lett.*, 92:120401, 2004.

[68] T. Bourdel, L. Khaykovich, J. Cubizolles, J. Zhang, F. Chevy, M. Teichmann, L. Tarruell, S. J. J. M. F. Kokkelmans, and C. Salomon. Experimental study of the BEC-BCS crossover region in Lithium 6. *Phys. Rev. Lett.*, 93:050401, 2004.

[69] J. R. Zwierlein, M. W. Abo-Shaer, A. Schirotzek, C. H. Schunck, and W. Ketterle. Vortices and superfluidity in a strongly interacting Fermi gas. *Nature*, 435:1047, 2005.

[70] M. W. Zwierlein, A. Schirotzek, C. H. Schunck, and W. Ketterle. Fermionic superfluidity with imbalanced spin populations. *Science*, 311:492, 2006.

[71] G. B. Partridge, W. Li, R. I. Kamar, Y. Liao, and R. G. Hulet. Pairing and phase separation in a polarized Fermi gas. *Science*, 311:503, 2006.

[72] J. Joseph, B. Clancy, L. Luo, J. Kinast, A. Turlapov, and J. E. Thomas. Measurement of sound velocity in a Fermi gas near a Feshbach resonance. *Phys. Rev. Lett.*, 98:170401, 2007.

[73] D. E. Miller, J. K. Chin, C. A. Stan, Y. Liu, W. Setiawan, C. Sanner, and W. Ketterle. Critical velocity for superfluid flow across the BEC-BCS crossover. *Phys. Rev. Lett.*, 99:070402, 2007.

[74] R. Jördens, K. Strohmaier, H. Günter, H. Moritz, and T. Esslinger. A Mott insulator of fermionic atoms in an optical lattice. *Nature*, 455:204, 2008.

[75] U. Schneider, L. Hackermüller, S. Will, T. Best, I. Bloch, T. A. Costi, R. W. Helmes, D. Rasch, and A. Rosch. Metallic and insulating phases of repulsively interacting Fermions in a 3D optical lattice. *Science*, 322:1520, 2008.

[76] M. Horikoshi, S. Nakajima, M. Ueda, and T. Mukaiyama. Measurement of universal thermodynamic functions for a unitary Fermi gas. *Science*, 327:442, 2010.

[77] N. Navon, S. Nascimbène, F. Chevy, and C. Salomon. The equation of state of a low-temperature Fermi gas with tunable interactions. *Science*, 328:729, 2010.

[78] P. Fulde and R. A. Ferrell. Superconductivity in a strong spin-exchange field. *Phys. Rev.*, 135:A550, 1964.

[79] A. I. Larkin and Y. N. Ovchinnikov. Inhomogeneous state of superconductors. *Sov. Phys. JETP*, 20:762, 1965.

[80] W. V. Liu and F. Wilczek. Interior gap superfluidity. *Phys. Rev. Lett.*, 90:047002, 2003.

[81] M. M. Forbes, E. Gubankova, W. V. Liu, and F. Wilczek. Stability criteria for breached-pair superfluidity. *Phys. Rev. Lett.*, 94:017001, 2005.

[82] D. S. Petrov, G. E. Astrakharchik, D. J. Papoular, C. Salomon, and G. V. Shlyapnikov. Crystalline phase of strongly interacting Fermi mixtures. *Phys. Rev. Lett.*, 99:130407, 2007.

[83] J. Levinsen, T. G. Tiecke, J. T. M. Walraven, and D. S. Petrov. Atom-dimer scattering and long-lived trimers in fermionic mixtures. *Phys. Rev. Lett.*, 103:153202, 2009.

[84] J. Deiglmayr, A. Grochola, M. Repp, K. Mörtlbauer, C. Glück, J. Lange, O. Dulieu, R. Wester, and M. Weidemüller. Formation of ultracold polar molecules in the rovibrational ground state. *Phys. Rev. Lett.*, 101:133004, 2008.

[85] K.-K. Ni, S. Ospelkaus, M. H. G. de Miranda, A. Pe'er, B. Neyenhuis, J. J. Zirbel, S. Kotochigova, P. S. Julienne, D. S. Jin, and J. Ye. A high phase-space-density gas of polar molecules. *Science*, 322:231, 2008.

[86] U. Gavish and Y. Castin. Matter-wave localization in disordered cold atom lattices. *Phys. Rev. Lett.*, 95:020401, 2005.

[87] Y. Nishida and S. Tan. Universal Fermi gases in mixed dimensions. *Phys. Rev. Lett.*, 101:170401, 2008.

[88] M. Aymar and O. Dulieu. Calculation of accurate permanent dipole moments of the lowest $^{1,3}\sigma^+$ states of heteronuclear alkali dimers using extended basis sets. *J. Chem. Phys.*, 122:204302, 2005.

[89] M. Taglieber, A.-C. Voigt, F. Henkel, S. Fray, T. W. Hänsch, and K. Dieckmann. Simultaneous magneto-optical trapping of three atomic species. *Phys. Rev. A*, 73:011402, 2006.

[90] E. Wille, F. M. Spiegelhalder, G. Kerner, D. Naik, A. Trenkwalder, G. Hendl, F. Schreck, R. Grimm, T. G. Tiecke, J. T. M. Walraven, S. J. J. M. F. Kokkelmans, E. Tiesinga, and P. S. Julienne. Exploring an ultracold Fermi-Fermi mixture: Interspecies Feshbach resonances and scattering properties of ^{6}Li and ^{40}K. *Phys. Rev. Lett.*, 100:053201, 2008.

[91] E. Tiemann, H. Knöckel, P. Kowalczyk, W. Jastrzebski, A. Pashov, H. Salami, and A. J. Ross. Coupled system $a\,^3\Sigma^+$ and $X\,^1\Sigma^+$ of KLi: Feshbach resonances and corrections to the Born-Oppenheimer approximation. *Phys. Rev. A*, 79:042716, 2009.

[92] T. G. Tiecke, M. R. Goosen, A. Ludewig, S. D. Gensemer, S. Kraft, S. J. J. M. F. Kokkelmans, and J. T. M. Walraven. Broad Feshbach resonance in the ^{6}Li-^{40}K mixture. *Phys. Rev. Lett.*, 104:053202, 2010.

[93] A.-C. Voigt, M. Taglieber, L. Costa, T. Aoki, W. Wieser, T. W. Hänsch, and K. Dieckmann. Ultracold heteronuclear Fermi-Fermi molecules. *Phys. Rev. Lett.*, 102:020405, 2009.

[94] K. M. O'Hara, S. L. Hemmer, M. E. Gehm, S. R. Granade, and J. E. Thomas. Observation of a strongly interacting degenerate Fermi gas of atoms. *Science*, 298:2179, 2002.

[95] A. Trenkwalder, C. Kohstall, M. Zaccanti, D. Naik, A. I. Sidorov, F. Schreck, and R. Grimm. Hydrodynamic expansion of a strongly interacting Fermi-Fermi mixture. *Phys. Rev. Lett.*, 106:115304, 2011.

[96] K. Góral, L. Santos, and M. Lewenstein. Quantum phases of dipolar bosons in optical lattices. *Phys. Rev. Lett.*, 88:170406, 2002.

[97] B. Damski, L. Santos, E. Tiemann, M. Lewenstein, S. Kotochigova, P. Julienne, and P. Zoller. Creation of a dipolar superfluid in optical lattices. *Phys. Rev. Lett.*, 90:110401, 2003.

[98] L. D. Carr, D. DeMille, R. V. Krems, and J. Ye. Cold and ultracold molecules: science, technology, and applications. *New J. Phys.*, 11:055049, 2009.

[99] R. V. Krems. Molecules near absolute zero and external field control of atomic and molecular dynamics. *Int. Rev. Phys. Chem.*, 24:99, 2007.

[100] D. DeMille. Quantum computation with trapped polar molecules. *Phys. Rev. Lett.*, 88:067901, 2002.

[101] P. Rabl, D. DeMille, J. M. Doyle, M. D. Lukin, R. J. Schoelkopf, and P. Zoller. Hybrid quantum processors: Molecular ensembles as quantum memory for solid state circuits. *Phys. Rev. Lett.*, 97:033003, 2006.

[102] M. G. Kozlov and D. DeMille. Enhancement of the electric dipole moment of the electron in PbO. *Phys. Rev. Lett.*, 89:133001, 2002.

[103] J. J. Hudson, B. E. Sauer, M. R. Tarbutt, and E. A. Hinds. Measurement of the electron electric dipole moment using YbF molecules. *Phys. Rev. Lett.*, 89:023003, 2002.

[104] D. DeMille, S. Sainis, J. Sage, T. Bergeman, S. Kotochigova, and E. Tiesinga. Enhanced sensitivity to variation of m_e/m_p in molecular spectra. *Phys. Rev. Lett.*, 100:043202, 2008.

[105] A. Shelkovnikov, R. J. Butcher, C. Chardonnet, and A. Amy-Klein. Stability of the proton-to-electron mass ratio. *Phys. Rev. Lett.*, 100:150801, 2008.

[106] E. R. Hudson, H. J. Lewandowski, B. C. Sawyer, and J. Ye. Cold molecule spectroscopy for constraining the evolution of the fine structure constant. *Phys. Rev. Lett.*, 96:143004, 2006.

[107] S. Yi and L. You. Trapped atomic condensates with anisotropic interactions. *Phys. Rev. A*, 61:041604, 2000.

[108] J. T. Bahns, W. C. Stwalley, and P. L. Gould. Laser cooling of molecules: A sequential scheme for rotation, translation, and vibration. *J. Chem. Phys.*, 104:9689, 1996.

[109] M. D. Di Rosa. Concept, candidates, and supporting hyperfine-resolved measurements of rotational lines in the A-X(0,0) band of CaH. *Eur. Phys. J. D*, 31:395, 2004.

[110] E. S. Shuman, J. F. Barry, and D. DeMille. Laser cooling of a diatomic molecule. *Nature*, 467:820, 2010.

[111] J. Weinstein, R. deCarvalho, T. Guillet, B. Friedrich, and J. M. Doyle. Magnetic trapping of calcium monohydride molecules at millikelvin temperatures. *Nature*, 395:148, 1998.

[112] H. L. Bethlem and G. Meijer. Production and application of translationally cold molecules. *Int. Rev. Phys. Chem.*, 22:73, 2003.

[113] S. A. Rangwala, T. Junglen, T. Rieger, P. W. H. Pinkse, and G. Rempe. Continuous source of translationally cold dipolar molecules. *Phys. Rev. A*, 67:043406, 2003.

[114] H. R. Thorsheim, J. Weiner, and P. S. Julienne. Laser-induced photoassociation of ultracold sodium atoms. *Phys. Rev. Lett.*, 58:2420, 1987.

[115] K. Bergmann, H. Theuer, and B. W. Shore. Coherent population transfer among quantum states of atoms and molecules. *Rev. Mod. Phys.*, 70:1003, 1998.

[116] U. Gaubatz, R. Rudecki, M. Becker, S. Schiemann, M. Külz, and K. Bergmann. Population switching between vibrational levels in molecular beams. *Chem. Phys. Lett.*, 149:463, 1988.

[117] J. R. Kuklinski, U. Gaubatz, F. T. Hioe, and K. Bergmann. Adiabatic population transfer in a three-level system driven by delayed laser pulses. *Phys. Rev. A*, 40:6741, 1989.

[118] J.G. Danzl, E. Haller, M. Gustavsson, M. J. Mark, R. Hart, N. Bouloufa, O. Dulieu, H. Ritsch, and H.-C. Nägerl. Quantum gas of deeply bound ground state molecules. *Science*, 321:1062, 2008.

[119] F. Lang, K. Winkler, C. Strauss, R. Grimm, and J. Hecker Denschlag. Ultracold triplet molecules in the rovibrational ground state. *Phys. Rev. Lett.*, 101:133005, 2008.

[120] P. S. Żuchowski and J. M. Hutson. Reactions of ultracold alkali-metal dimers. *Phys. Rev. A*, 81:060703, 2010.

[121] M. H. G. de Miranda, A. Chotia, B. Neyenhuis, D. Wang, G. Quemener, S. Ospelkaus, J. L. Bohn, J. Ye, and D. S. Jin. Controlling the quantum stereodynamics of ultracold bimolecular reactions. *Nature Phys.*, 2011.

[122] A. Ridinger, S. Chaudhuri, T. Salez, U. Eismann, D. R. Fernandes, K. Magalhães, D. Wilkowski, C. Salomon, and F. Chevy. Large atom number dual-species magneto-optical trap for fermionic ^{6}Li and ^{40}K atoms. *Eur. Phys. J. D, DOI:10.1140/epjd/e2011-20069-4*, 2011.

[123] A. Ridinger, S. Chaudhuri, T. Salez, D. R. Fernandes, N. Bouloufa, O. Dulieu, C. Salomon, and F. Chevy. Photoassociative creation of ultracold heteronuclear ^{6}Li^{40}K* molecules. *Eur. Phys. Lett., in press*, 2011.

[124] A. Ridinger and N. Davidson. Particle motion in rapidly oscillating potentials: The role of the potential's initial phase. *Phys. Rev. A*, 76:013421, 2007.

[125] A. Ridinger and C. Weiss. Manipulation of quantum particles in rapidly oscillating potentials by inducing phase hops. *Phys. Rev. A*, 79:013414, 2009.

[126] P. W. Cleary, T. W. Hijmans, and J. T. M. Walraven. Manipulation of a Bose-Einstein condensate by a time-averaged orbiting potential using phase jumps of the rotating field. *Phys. Rev. A*, 82:063635, 2010.

[127] K. L. Moore, T. P. Purdy, K. W. Murch, S. Leslie, S. Gupta, and D. M. Stamper-Kurn. Collimated, single-pass atom source from a pulsed alkali metal dispenser for laser-cooling experiments. *Rev. Sci. Instrum.*, 76:023106, 2005.

[128] A. Gozzini, F. Mango, J. H. Xu, G. Alzetta, F. Maccarrone, and R. A. Bernheim. Light-induced ejection of alkali atoms in polysiloxane coated cells. *Nuovo Cimento D*, 15:709, 1993.

[129] C. Klempt, T. van Zoest, T. Henninger, O. Topic, E. Rasel, W. Ertmer, and J. Arlt. Ultraviolet light-induced atom desorption for large rubidium and potassium magneto-optical traps. *Phys. Rev. A*, 73:013410, 2006.

[130] T. L. Gustavson, A. P. Chikkatur, A. E. Leanhardt, A. Görlitz, S. Gupta, D. E. Pritchard, and W. Ketterle. Transport of Bose-Einstein condensates with optical tweezers. *Phys. Rev. Lett.*, 88:020401, 2001.

[131] M. Greiner, I. Bloch, T. W. Hänsch, and T. Esslinger. Magnetic transport of trapped cold atoms over a large distance. *Phys. Rev. A*, 63:031401, 2001.

[132] G. Ferrari, M.-O. Mewes, F. Schreck, and C. Salomon. High-power multiple-frequency narrow-linewidth laser source based on a semiconductor tapered amplifier. *Opt. Lett.*, 24:151, 1999.

[133] K. A. Yakimovich and A. G. Mozgovoi. Experimental investigation of the density and surface tension of molten lithium at temperatures up to 1300 k. *High Temp.*, 38:657, 2000.

[134] W. D. Phillips and H. Metcalf. Laser deceleration of an atomic beam. *Phys. Rev. Lett.*, 48:596, 1982.

[135] H. J. Metcalf and P. van der Straten. *Laser cooling and trapping*. Springer, Berlin, 1999.

[136] F. Schreck. *Ph.D. thesis*. Ecole Normale Supérieure, Paris, 2002.

[137] J. H. Moore, C. C. Davis, and M. A. Coplan. *Building Scientific Apparatus*. Westview Press, Boulder, Colorado, 2002.

[138] N. F. Ramsey. *Molecular Beams*. Oxford University Press, Oxford, 1986.

[139] A. Joffe, W. Ketterle, A. Martin, and D. E. Pritchard. Transverse cooling and deflection of an atomic beam inside a zeeman slower. *J. Opt. Soc. Am. B*, 10:2257, 1993.

[140] K. Dieckmann, R. J. C. Spreeuw, M. Weidemüller, and J. T. M. Walraven. Two-dimensional magneto-optical trap as a source of slow atoms. *Phys. Rev. A*, 58:3891, 1998.

[141] J. Schoser, A. Batär, R. Löw, V. Schweikhard, A. Grabowski, Yu. B. Ovchinnikov, and T. Pfau. Intense source of cold Rb atoms from a pure two-dimensional magneto-optical trap. *Phys. Rev. A*, 66:023410, 2002.

[142] J. Catani, P. Maioli, L. De Sarlo, F. Minardi, and M. Inguscio. Intense slow beams of bosonic potassium isotopes. *Phys. Rev. A*, 73:033415, 2006.

[143] S. Chaudhuri, S. Roy, and C. S. Unnikrishnan. Realization of an intense cold Rb atomic beam based on a two-dimensional magneto-optical trap: Experiments and comparison with simulations. *Phys. Rev. A*, 74:023406, 2006.

[144] T. G. Tiecke, S. D. Gensemer, A. Ludewig, and J. T. M. Walraven. High-flux two-dimensional magneto-optical-trap source for cold lithium atoms. *Phys. Rev. A*, 80:013409, 2009.

[145] M. S. Santos, P. Nussenzveig, L. G. Marcassa, K. Helmerson, J. Flemming, S. C. Zilio, and V. S. Bagnato. Simultaneous trapping of two different atomic species in a vapor-cell magneto-optical trap. *Phys. Rev. A*, 52:R4340, 1995.

[146] U. Schlöder, H. Engler, U. Schünemann, R. Grimm, and M. Weidemüller. Cold inelastic collisions between lithium and cesium in a two-species magneto-optical trap. *Eur. Phys. J. D*, 7:331, 1999.

[147] G. D. Telles, W. Garcia, L. G. Marcassa, V. S. Bagnato, D. Ciampini, M. Fazzi, J. H. Müller, D. Wilkowski, and E. Arimondo. Trap loss in a two-species Rb-Cs magneto-optical trap. *Phys. Rev. A*, 63:033406, 2001.

[148] J. Goldwin, S. B. Papp, B. DeMarco, and D. S. Jin. Two-species magneto-optical trap with ^{40}K and ^{87}Rb. *Phys. Rev. A*, 65:021402, 2002.

[149] A. L. Migdall, J. V. Prodan, W. D. Phillips, T. H. Bergeman, and H. J. Metcalf. First observation of magnetically trapped neutral atoms. *Phys. Rev. Lett.*, 54:2596, 1985.

[150] W. Ketterle and D. E. Pritchard. Trapping and focusing ground state atoms with static fields. *Appl. Phys. B*, 54:403, 1982.

[151] G. Breit and I. I. Rabi. Measurement of nuclear spin. *Phys. Rev.*, 38:2082, 1931.

[152] W. Petrich, M. H. Anderson, J. R. Ensher, and E. A. Cornell. Stable, tightly confining magnetic trap for evaporative cooling of neutral atoms. *Phys. Rev. Lett.*, 74:3352, 1995.

[153] D. E. Pritchard. Cooling neutral atoms in a magnetic trap for precision spectroscopy. *Phys. Rev. Lett.*, 51:1336, 1983.

[154] K. B. Davis, M. O. Mewes, M. R. Andrews, N. J. van Druten, D. S. Durfee, D. M. Kurn, and W. Ketterle. Bose-Einstein condensation in a gas of sodium atoms. *Phys. Rev. Lett.*, 75:3969, 1995.

[155] D. L. Whitaker, H. J. Lewandowski, D. M. Harber, and E. A. Cornell. Simplified system for creating a Bose-Einstein condensate. *J. Low Temp. Phys.*, 132:309, 2003.

[156] T. Salez. *Ph.D. thesis*. Ecole Normale Supérieure, Paris, 2011.

[157] W. Ketterle and M. Zwierlein. Making, probing and understanding ultracold Fermi gases. In M. Inguscio, W. Ketterle, and C. Salomon, editors, *Proceedings of the International School of Physics Enrico Fermi on Ultracold Fermi Gases*, page 95, 2006.

[158] S. Falke, H. Knöckel, J. Friebe, M. Riedmann, E. Tiemann, and C. Lisdat. Potassium ground-state scattering parameters and Born-Oppenheimer potentials from molecular spectroscopy. *Phys. Rev. A*, 78:012503, 2008.

[159] V. Dribinski, A. Ossadtchi, V. A. Mandelshtam, and H. Reisler. Reconstruction of Abel-transformable images: The Gaussian basis-set expansion Abel transform method. *Rev. Sci. Instrum.*, 73:2634, 2002.

[160] A. D. Poularikas, editor. *The Transforms and Applications Handbook, 2nd ed.* CRC Press, Boca Raton, Florida, 2000.

[161] M. Teichmann. *Ph.D. thesis*. Ecole Normale Supérieure, Paris, 2007.

[162] C. B. Alcock, V. P. Itkin, and M. K. Horrigan. Vapour pressure equations for the metallic elements: 298-2500K. *Can. Metall. Q.*, 23:309, 1984.

[163] A. M. Steane, M. Chowdhury, and C. J. Foot. Radiation force in the magneto-optical trap. *J. Opt. Soc. Am. B*, 9:2142, 1992.

[164] A. Derevianko, W. R. Johnson, M. S. Safronova, and J. F. Babb. Radiation force in the magneto-optical trap. *Phys. Rev. Lett.*, 82:3589, 2002.

[165] M. Weidemüller and C. Zimmermann. *Interactions in Ultracold Gases*. Wiley-VCH, Weinheim, 2003.

[166] B. Bussery, Y. Achkar, and M. Aubert-Frécon. Long-range molecular states dissociating to the three of four lowest asymptotes for the ten heteronuclear diatomic alkali molecules. *Chem. Phys.*, 116:319, 1987.

[167] M. Marinescu and H. R. Sadeghpour. Long-range potentials for two-species alkali-metal atoms. *Phys. Rev. A*, 59:390, 1999.

[168] S. D. Gensemer, V. Sanchez-Villicana, K. Y. N. Tan, T. T. Grove, and P. L. Gould. Trap-loss collisions of ^{85}Rb and ^{87}Rb: Dependence on trap parameters. *Phys. Rev. A*, 56:4055, 1997.

[169] D. E. Fagnan, J. Wang, C. Zhu, P. Djuricanin, B. G. Klappauf, J. L. Booth, and K. W. Madison. Observation of quantum diffractive collisions using shallow atomic traps. *Phys. Rev. A*, 80:022712, 2009.

[170] P. D. Lett, P. S. Julienne, and W. D. Philipps. Photoassociative spectroscopy of laser-cooled atoms. *Annu. Rev. Phys. Chem.*, 46:423, 1995.

[171] W. C. Stwalley and H. Wang. Photoassociation of ultracold atoms: A new spectroscopic technique. *J. Mol. Spec.*, 195:194, 1999.

[172] F. Masnou-Seeuws and P. Pillet. Formation of ultracold molecules ($T \leq 200\,\mu K$) via photoassociation in a gas of laser-cooled atoms. *Adv. At. Mol. Opt. Phys.*, 47:53, 2001.

[173] K. M. Jones, E. Tiesinga, P. D. Lett, and P. S. Julienne. Ultracold photoassociation spectroscopy: Long-range molecules and atomic scattering. *Rev. Mod. Phys.*, 78:483, 2006.

[174] N. Bouloufa, A. Crubellier, and O. Dulieu. Photoassociative molecular spectroscopy for atomic radiative lifetimes. *Phys. Scr.*, T134:014014, 2008.

[175] L. D. Landau and E. M. Lifschitz. *Quantum Mechanics*. Butterworth-Heinemann, Oxford, UK, 1977.

[176] T. Köhler, K. Góral, and P. S. Julienne. Production of cold molecules via magnetically tunable Feshbach resonances. *Rev. Mod. Phys.*, 78:1311, 2006.

[177] E. R. I. Abraham, W. I. McAlexander, C. A Sackett, and R. G. Hulet. Spectroscopic determination of the s-wave scattering length of lithium. *Phys. Rev. Lett.*, 74:1315, 1995.

[178] R. Côté, A. Dalgarno, Y. Sun, and R. G. Hulet. Photoabsorption by ultracold atoms and the scattering length. *Phys. Rev. Lett.*, 74:3581, 1995.

[179] E. Tiesinga, C. J. Williams, P. S. Julienne, K. M. Jones, P. D. Lett, and W. D. Phillips. A spectroscopic determination of scattering lengths for sodium atom collisions. *J. Res. Natl. Inst. Stand. Technol.*, 101:505, 1996.

[180] J. R. Gardner, R. A. Cline, J. D. Miller, D. J. Heinzen, H. M. J. M. Boesten, and B. J. Verhaar. Collisions of doubly spin-polarized, ultracold ^{85}Rb atoms. *Phys. Rev. Lett.*, 74:3764, 1995.

[181] C. Drag, B. Laburthe Tolra, B. T'Jampens, D. Comparat, M. Allegrini, A. Crubellier, and P. Pillet. Photoassociative spectroscopy as a self-sufficient tool for the determination of the cs triplet scattering length. *Phys. Rev. Lett.*, 85:1408, 2000.

[182] Y. Takasu, K. Komori, K. Honda, M. Kumakura, T. Yabuzaki, and Y. Takahashi. Photoassociation spectroscopy of laser-cooled ytterbium atoms. *Phys. Rev. Lett.*, 93:123202, 2004.

[183] O. Dulieu and C. Gabbanini. The formation and interactions of cold and ultracold molecules: new challenges for interdisciplinary physics. *Rep. Prog. Phys.*, 72:086401, 2009.

[184] C. Chin, R. Grimm, P. Julienne, and E. Tiesinga. Feshbach resonances in ultracold gases. *Rev. Mod. Phys.*, 82:1225, 2010.

[185] P. D. Lett, K. Helmerson, W. D. Phillips, L. P. Ratliff, S. L. Rolston, and M. E. Wagshul. Spectroscopy of Na_2 by photoassociation of laser-cooled Na. *Phys. Rev. Lett.*, 71:2200, 1993.

[186] H. Wang, P. L. Gould, and W. C. Stwalley. Photoassociative spectroscopy of ultracold ^{39}K atoms in a high-density vapor-cell magneto-optical trap. *Phys. Rev. A*, 53:R1216, 1996.

[187] J. D. Miller, R. A. Cline, and D. J. Heinzen. Photoassociation spectrum of ultracold Rb atoms. *Phys. Rev. Lett.*, 71:2204, 1993.

[188] A. Fioretti, D. Comparat, A. Crubellier, O. Dulieu, F. Masnou-Seeuws, and P. Pillet. Formation of cold Cs_2 molecules through photoassociation. *Phys. Rev. Lett.*, 80:4402, 1998.

[189] A. P. Mosk, M. W. Reynolds, T. W. Hijmans, and J. T. M. Walraven. Photoassociation of spin-polarized hydrogen. *Phys. Rev. Lett.*, 82:307, 1999.

[190] N. Herschbach, P. J. J. Tol, W. Vassen, W. Hogervorst, G. Woestenenk, J. W. Thomsen, P. van der Straten, and A. Niehaus. Photoassociation spectroscopy of cold He(2^3S) atoms. *Phys. Rev. Lett.*, 84:1874, 2000.

[191] G. Zinner, T. Binnewies, F. Riehle, and E. Tiemann. Photoassociation of cold Ca atoms. *Phys. Rev. Lett.*, 85:2292, 2000.

[192] S. B. Nagel, P. G. Mickelson, A. D. Saenz, Y. N. Martinez, Y. C. Chen, T. C. Killian, P. Pellegrini, and R. Côté. Photoassociative spectroscopy at long range in ultracold strontium. *Phys. Rev. Lett.*, 94:083004, 2005.

[193] U. Schlöder, C. Silber, T. Deuschle, and C. Zimmermann. Saturation in heteronuclear photoassociation of ^{6}Li^{7}Li. *Phys. Rev. A*, 66:061403, 2002.

[194] A. J. Kerman, J. M. Sage, S. Sainis, T. Bergeman, and D. DeMille. Production of ultracold, polar RbCs* molecules via photoassociation. *Phys. Rev. Lett.*, 92:033004, 2004.

[195] D. Wang, J. Qi, M. F. Stone, O. Nikolayeva, H. Wang, B. Hattaway, S. D. Gensemer, P. L. Gould, E. E. Eyler, and W. C. Stwalley. Photoassociative production and trapping of ultracold KRb molecules. *Phys. Rev. Lett.*, 93:243005, 2004.

[196] C. Haimberger, J. Kleinert, M. Bhattacharya, and N. P. Bigelow. Formation and detection of ultracold ground-state polar molecules. *Phys. Rev. A*, 70:021402, 2004.

[197] N. Nemitz, F. Baumer, F. Münchow, S. Tassy, and A. Görlitz. Production of heteronuclear molecules in an electronically excited state by photoassociation in a mixture of ultracold Yb and Rb. *Phys. Rev. A*, 79:061403, 2009.

[198] H. Wang and W. C. Stwalley. Ultracold photoassociative spectroscopy of heteronuclear alkali-metal diatomic molecules. *J. Chem. Phys.*, 108:5767, 1998.

[199] S. Azizi, M. Aymar, and O. Dulieu. Prospects for the formation of ultracold ground state polar molecules from mixed alkali atom pairs. *Eur. Phys. J. D*, 31:195, 2004.

[200] H. Wang, P. L. Gould, and W. C. Stwalley. Fine-structure predissociation of ultracold photoassociated $^{39}K_2$ molecules observed by fragmentation spectroscopy. *Phys. Rev. Lett.*, 80:476, 1998.

[201] J. Léonard, A. P. Mosk, M. Walhout, P. van der Straten, M. Leduc, and C. Cohen-Tannoudji. Analysis of photoassociation spectra for giant helium dimers. *Phys. Rev. A*, 69:032702, 2004.

[202] R. V. Krems, W. C. Stwalley, and F. Bretislav, editors. *Cold Molecules: Theory, experiment, applications.* CRC Press, Boca Raton, Florida, 2009.

[203] M. Movre and R. Beuc. Van der Waals interaction in excited alkali-metal dimers. *Phys. Rev. A*, 31:2957, 1985.

[204] M. Marinescu and A. Dalgarno. Dispersion forces and long-range electronic transition dipole moments of alkali-metal dimer excited states. *Phys. Rev. A*, 52:311, 1995.

[205] P. S. Julienne and J. Vigué. Cold collisions of ground- and excited-state alkali-metal atoms. *Phys. Rev. A*, 44:4464, 1991.

[206] M. Aymar, N. Bouloufa, and O. Dulieu. private communication.

[207] G. Hertzberg, editor. *Molecular Spectra and Molecular Structure: I. Spectra of Diatomic Molecules, 2nd ed.* van Nostrand Reinhold, Princeton, New York, 1950.

[208] J. R. LeRoy and R. B. Bernstein. Dissociation energy and long-range potential of diatomic molecules from vibrational spacings of higher levels. *J. Chem. Phys.*, 52:3869, 1970.

[209] W. C. Stwalley. The dissociation energy of the hydrogen molecule using long-range forces. *Chem. Phys. Lett.*, 6:241, 1970.

[210] B. Ji, C. C. Tsai, and W. C. Stwalley. Proposed modification of the criterion for the region of validity of the inverse-power expansion in diatomic long-range potentials. *Chem. Phys. Lett.*, 236:242, 1995.

[211] C. Boisseau, Audouard E., and J. Vigué. Quantization of the highest levels in a molecular potential. *Europhys. Lett.*, 41:349, 1998.

[212] D. Comparat. Improved LeRoy-Bernstein near-dissociation expansion formula and prospect for photoassociation spectroscopy. *J. Chem. Phys.*, 120:1318, 2003.

[213] F. Engelke, H. Hage, and U. Sprick. The $B^1\Pi - X^1\Sigma^+$ system of $^{39}K^6Li$ and $^{39}K^7Li$: High-resolution laser excitation and fluorescence spectroscopy using selectively detected laser-induced fluorescence (SDLIF) in molecular beam and injection heat pipe (IHP). *Chem. Phys.*, 88:443, 1984.

[214] H. Salami, A. Ross, J. Crozet, W. Jastrzebski, P. Kowalczyk, and R. J. LeRoy. A full analytical energy curve for the $a^3\sigma^+$ state of kli from a limited vibrational data set. *J. Chem. Phys.*, 126:194313, 2007.

[215] A. Pashov, W. Jastrzebski, and P. Kowalczyk. The $B^1\Pi$ and $C^1\Sigma^+$ states of KLi. *Chem. Phys. Lett.*, 292:615, 1998.

[216] W. Jastrzebski, P. Kowalczyk, and A. Pashov. The perturbation of the $B^1\Pi$ and $C^1\Sigma^+$ states in KLi. *J. Mol. Spectr.*, 209:50, 2001.

[217] A. Grochola, W. Jastrzebski, P. Kowalczyk, P. Crozet, and A. J. Ross. The molecular constants and potential energy curve of the $D^1\Pi$ state in KLi. *Chem. Phys. Lett.*, 372:173, 2003.

[218] A. Grochola, W. Jastrzebski, P. Kortyka, and P. Kowalczyk. Polarization labelling spectroscopy of the $4^1\Pi$ state of KLi. *Mol. Phys.*, 102:1739, 2004.

[219] A. Grochola, A. Pashov, J. Deiglmayr, M. Repp, E. Tiemann, R. Wester, and M. Weidemüller. Photoassociation spectroscopy of the $B^1\Pi$ state of LiCs. *J. Chem. Phys.*, 131:054304, 2009.

[220] T. Führer and T. Walther. Extension of the mode-hop-free tuning range of an external cavity diode laser based on a model of the mode-hop dynamics. *Opt. Lett.*, 33:372, 2008.

[221] R. Grimm, M. Weidemüller, and Y. B. Ovchinnikov. Optical dipole traps for neutral atoms. *Ad. At. Mol. Opt. Phys.*, 42:95, 2000.

[222] W. C. Stwalley, Y.-H. Uang, and G. Pichler. Pure long-range molecules. *Phys. Rev. Lett.*, 41:1164, 1978.

[223] B. Bussery and M. Aubert-Frécon. Potential energy curves and vibration-rotation energies for the two purely long-range bound states 1_u and 0_g^- of the alkali dimers M_2 dissociating to $M(n^s 2S_{1/2}) + M(n^p 2P_{3/2})$ with M = Na, K, Rb, and Cs. *J. Mol. Spec.*, 113:21, 1985.

[224] H. Wang, P. L. Gould, and W. C. Stwalley. Long-range interaction of the $^{39}K(4s)+^{39}K(4p)$ asymptote by photoassociative spectroscopy. I. The 0_g^- pure long-range state and the long-range potential constants. *J. Chem. Phys.*, 106:7899, 1997.

[225] H. Wang, J. Li, X. T. Wang, C. J. Williams, P. L. Gould, and W. C. Stwalley. Precise determination of the dipole matrix element and radiative lifetime of the ^{39}K 4p state by photoassociative spectroscopy. *Phys. Rev. A*, 55:R1569, 1997.

[226] http://physics.nist.gov/physrefdata/asd/lines_form.html.

[227] V. Bednarska, A. Ekers, P. Kowalczyk, and W. Jastrzebski. Doppler-free spectroscopy of KLi. *J. Chem. Phys.*, 106:6332, 1997.

[228] C. J. Williams, E. Tiesinga, and P. S. Julienne. Hyperfine structure of the $Na_2\ 0_g^-$ long-range molecular state. *Phys. Rev. A*, 53:R1939, 1996.

[229] P. Pillet, A. Crubellier, A. Bleton, O. Dulieu, P. Nosbaum, I. Mourachko, and F. Masnou-Seeuws. Photoassociation in a gas of cold alkali atoms: I. Perturbative quantum approach. *J. Phys. B*, 30:2801, 1997.

[230] J. L. Bohn and P. S. Julienne. Semianalytic theory of laser-assisted resonant cold collisions. *Phys. Rev. A*, 60:414, 1999.

[231] J. Javanainen and M. Mackie. Probability of photoassociation from a quasicontinuum approach. *Phys. Rev. A*, 58:R789, 1998.

[232] P. L. Kapitza. Dinamicheskaya ustoichivost mayatnika pri koleblyushcheisya tochke podvesa. *Zh. Eksp. Teor. Fiz.*, 21:588, 1951.

[233] W. Paul. Electromagnetic traps for charged and neutral particles. *Rev. Mod. Phys.*, 62:531, 1990.

[234] D. Leibfried, R. Blatt, C. Monroe, and D. Wineland. Quantum dynamics of single trapped ions. *Rev. Mod. Phys.*, 75:281, 2003.

[235] F. G. Major, V. N. Gheorghe, and G. Werth. *Charged Particle Traps: Physics and Techniques of Charged Particle Field Confinement*. Springer-Verlag, New York, 2004.

[236] T. Junglen, T. Rieger, S. A. Rangwala, P. W. H. Pinkse, and G. Rempe. Two-dimensional trapping of dipolar molecules in time-varying electric fields. *Phys. Rev. Lett.*, 92:223001, 2004.

[237] J. van Veldhoven, H. L. Bethlem, and G. Meijer. ac electric trap for ground-state molecules. *Phys. Rev. Lett.*, 94:083001, 2005.

[238] R. V. E. Lovelace, C. Mehanian, T. J. Tommila, and D. M. Lee. Magnetic confinement of a neutral gas. *Nature*, 318:30, 1985.

[239] E. A. Cornell, C. Monroe, and C. E. Wieman. Multiply loaded, ac magnetic trap for neutral atoms. *Phys. Rev. Lett.*, 67:2439, 1991.

[240] T. Kishimoto, H. Hachisu, J. Fujiki, K. Nagato, M. Yasuda, and H. Katori. Electrodynamic trapping of spinless neutral atoms with an atom chip. *Phys. Rev. Lett.*, 96:123001, 2006.

[241] S. Schlunk, A. Marian, P. Geng, A. P. Mosk, G. Meijer, and W. Schöllkopf. Trapping of Rb atoms by ac electric fields. *Phys. Rev. Lett.*, 98:223002, 2007.

[242] M. Edwards, R. J. Dodd, C. W. Clark, P. A. Ruprecht, and K. Burnett. Properties of a Bose-Einstein condensate in an anisotropic harmonic potential. *Phys. Rev. A*, 53:R1950, 1996.

[243] V. I. Yukalov. Nonadiabatic dynamics of atoms in nonuniform magnetic fields. *Phys. Rev. A*, 56:5004, 1997.

[244] V. G. Minogin, J. A. Richmond, and G. I. Opat. Time-orbiting-potential quadrupole magnetic trap for cold atoms. *Phys. Rev. A*, 58:3138, 1998.

[245] D. J. Han, R. H. Wynar, Ph. Courteille, and D. J. Heinzen. Bose-Einstein condensation of large numbers of atoms in a magnetic time-averaged orbiting potential trap. *Phys. Rev. A*, 57:R4114, 1998.

BIBLIOGRAPHY

[246] B. P. Anderson and M. A. Kasevich. Spatial observation of Bose-Einstein condensation of ^{87}Rb in a confining potential. *Phys. Rev. A*, 59:R938, 1999.

[247] R. Franzosi, B. Zambon, and E. Arimondo. Nonadiabatic effects in the dynamics of atoms confined in a cylindric time-orbiting-potential magnetic trap. *Phys. Rev. A*, 70:053603, 2004.

[248] V. Milner, J. L. Hanssen, W. C. Campbell, and M. G. Raizen. Optical billiards for atoms. *Phys. Rev. Lett.*, 86:1514, 2001.

[249] N. Friedman, A. Kaplan, D. Carasso, and N. Davidson. Observation of chaotic and regular dynamics in atom-optics billiards. *Phys. Rev. Lett.*, 86:1518, 2001.

[250] P. Ahmadi, B. P. Timmons, and G. S. Summy. Geometrical effects in the loading of an optical atom trap. *Phys. Rev. A*, 72:023411, 2005.

[251] S. K. Schnelle, E. D. van Ooijen, M. J. Davis, N. R. Heckenberg, and H. Rubinsztein-Dunlop. Versatile two-dimensional potentials for ultra-cold atoms. *Opt. Express*, 16:1405, 2008.

[252] K. Sasaki, M. Koshioka, H. Misawa, N. Kitamura, and H. Masuhara. Pattern formation and flow control of fine particles by laser-scanning micromanipulation. *Opt. Lett.*, 16:1463, 1991.

[253] K. C. Neuman and S. M. Block. Optical trapping. *Rev. Sci. Instrum.*, 75:2787, 2004.

[254] L. D. Landau and E. M. Lifschitz. *Mechanics*. Pergamon Press, Oxford, UK, 1976.

[255] R. J. Cook, D. G. Shankland, and A. L. Wells. Quantum theory of particle motion in a rapidly oscillating field. *Phys. Rev. A*, 31:564, 1985.

[256] T. P. Grozdanov and M. J. Rakovic. Quantum system driven by rapidly varying periodic perturbation. *Phys. Rev. A*, 38:1739, 1988.

[257] S. Rahav, I. Gilary, and S. Fishman. Time independent description of rapidly oscillating potentials. *Phys. Rev. Lett.*, 91:110404, 2003.

[258] S. Rahav, I. Gilary, and S. Fishman. Effective Hamiltonians for periodically driven systems. *Phys. Rev. A*, 68:013820, 2003.

[259] J. A. Sanders and F. Verhulst. *Averaging Methods in Nonlinear Dynamical Systems*. Springer-Verlag, New York, 1985.

[260] F. G. Major and H. G. Dehmelt. Exchange-collision technique for the rf spectroscopy of stored ions. *Phys. Rev.*, 170:91, 1968.

[261] J. H. Shirley. Solution of the Schrödinger equation with a Hamiltonian periodic in time. *Phys. Rev.*, 138:B979, 1965.

BIBLIOGRAPHY 221

[262] Y. B. Zel'dovich. The quasienergy of a quantum-mechanical system subjected to a periodic action. *Sov. Phys. JETP*, 24:1006, 1967.

[263] N. Gross and L. Khaykovich. All-optical production of ^{7}Li Bose-Einstein condensation using Feshbach resonances. *Phys. Rev. A*, 77:023604, 2008.

[264] J. E. Thomas, J. Kinast, and A. Turlapov. Virial theorem and universality in a unitary Fermi gas. *Phys. Rev. Lett.*, 95:120402, 2005.

[265] E. Schrödinger. Der stetige übergang von der Mikro- zur Makromechanik. *Naturwissenschaften*, 14:664, 1926.

[266] P. Carruthers and M. M. Nieto. Coherent states and the forced quantum oscillator. *Am. J. Phys*, 33:537, 1965.

[267] F. Grossmann, T. Dittrich, P. Jung, and P. Hänggi. Coherent destruction of tunneling. *Phys. Rev. Lett.*, 67:516, 1991.

[268] E. Kierig, U. Schnorrberger, A. Schietinger, J. Tomkovic, and M. K. Oberthaler. Single-particle tunneling in strongly driven double-well potentials. *Phys. Rev. Lett.*, 100:190405, 2008.

[269] A. Eckardt, T. Jinasundera, C. Weiss, and M. Holthaus. Analog of photon-assisted tunneling in a Bose-Einstein condensate. *Phys. Rev. Lett.*, 95:200401, 2005.

[270] C. Sias, H. Lignier, Y. P. Singh, A. Zenesini, D. Ciampini, O. Morsch, and E. Arimondo. Observation of photon-assisted tunneling in optical lattices. *Phys. Rev. Lett.*, 100:040404, 2008.

[271] H. Lignier, C. Sias, D. Ciampini, Y. Singh, A. Zenesini, O. Morsch, and E. Arimondo. Dynamical control of matter-wave tunneling in periodic potentials. *Phys. Rev. Lett.*, 99:220403, 2007.

[272] A. Alberti, V. V. Ivanov, G. M. Tino, and G. Ferrari. Engineering the quantum transport of atomic wavefunctions over macroscopic distances. *Nat. Phys.*, 5:547, 2009.

[273] E. Haller, R. Hart, M. J. Mark, J. G. Danzl, L. Reichsöllner, and H.-C. Nägerl. Inducing transport in a dissipation-free lattice with super bloch oscillations. *Phys. Rev. Lett.*, 104:200403, 2010.

[274] K. Kudo and T. S. Monteiro. Theoretical analysis of super-Bloch oscillations. *Phys. Rev. A*, 83:053627, 2011.

[275] R. Casalbuoni and G. Nardulli. Inhomogeneous superconductivity in condensed matter and QCD. *Rev. Mod. Phys.*, 76:263, 2004.

[276] C. Chin, V. V. Flambaum, and M. G. Kozlov. Ultracold atoms: new probes on the variation of fundamental constants. *New J. Phys.*, 11:055048, 2009.

[277] S. Falke, E. Tiemann, C. Lisdat, H. Schnatz, and G. Grosche. Transition frequencies of the D lines of ^{39}K, ^{40}K, and ^{41}K measured with a femtosecond laser frequency comb. *Phys. Rev. A*, 74:032503, 2006.

[278] T. Tiecke. Properties of potassium. 2010.

[279] W. Kuhn. Über die Gesamtstärke der von einem Zustande ausgehenden Absorptionslinien. *Z. Phys.*, 33:408, 1925.

[280] F. Reiche and W. Thomas. Über die Zahl der Dispersionelektronen die einem stationären Zustand zugeordnet sind. *Z. Phys.*, 34:510, 1925.

[281] A. M. Mazouchi. *Master thesis*. University of Toronto, 2007.

[282] A. Hemmerich, D. H. McIntyre, C. Zimmermann, and T. W. Hänsch. Second-harmonic generation and optical stabilization of a diode laser in an external ring resonator. *Opt. Lett.*, 15:372, 1990.

[283] N. Ito. Doppler-free modulation transfer spectroscopy of rubidium $5^2S_{1/2}-6^2P_{1/2}$ transitions using a frequency-doubled diode laser blue-light source. *Rev. Sci. Inst.*, 71:2655, 1999.

[284] K. Hayasaka. Frequency stabilization of an extended-cavity violet diode laser by resonant optical feedback. *Opt. Comm.*, 206:401, 2002.

[285] U. Gustafsson, J. Alnis, and S. Svanberg. Atomic spectroscopy with violet laser diodes. *Am. J. Phys.*, 68:660, 2000.

[286] S. Uetake, K. Hayasaka, and M. Watanabe. Saturation spectroscopy of potassium for frequency stabilization of violet diode lasers. *Jpn. J. Appl. Phys.*, 42:L332, 2003.

[287] A. Behrle, M. Koschorreck, and M. Köhl. Isotope shift and hyperfine splitting of the $4S \to 5P$ transition in potassium. *Phys. Rev. A*, 83:052507, 2011.

[288] A. Reed. *Senior honors thesis*. Ohio State University, 2011.

[289] D.F. Walls and G.J. Milburn. *Quantum Optics, 2nd edition*. Springer Verlag, Berlin Heidelberg, 2002.

[290] J. Wang, L. B. Kong, K. J. Jiang, K. Li, X. H. Tu, H. W. Xiong, Y. Zhu, and M. S. Zhan. Electromagnetically induced transparency in multi-level cascade scheme of cold rubidium atoms. *Phys. Lett. A*, 328:437, 2004.

[291] A. Lindgard and S. E. Nielsen. Transition probabilities for the alkali isoelectronic sequences Li I, Na I, K I, Rb I, Cs I, Fr I. *At. Data Nucl. Data Tables*, 19:533, 1997.

Die VDM Verlagsservicegesellschaft sucht für wissenschaftliche Verlage abgeschlossene und herausragende

Dissertationen, Habilitationen, Diplomarbeiten, Master Theses, Magisterarbeiten usw.

für die kostenlose Publikation als Fachbuch.

Sie verfügen über eine Arbeit, die hohen inhaltlichen und formalen Ansprüchen genügt, und haben Interesse an einer honorarvergüteten Publikation?

Dann senden Sie bitte erste Informationen über sich und Ihre Arbeit per Email an *info@vdm-vsg.de*.

Sie erhalten kurzfristig unser Feedback!

VDM Verlagsservicegesellschaft mbH
Dudweiler Landstr. 99
D - 66123 Saarbrücken

Telefon +49 681 3720 174
Fax +49 681 3720 1749

www.vdm-vsg.de

Die VDM Verlagsservicegesellschaft mbH vertritt

Printed by Books on Demand GmbH, Norderstedt / Germany